Burning for Success

Burning for Success

How Volunteer Fire Departments Motivate Teams, Coach Leaders and Deliver Killer Customer Service Without Spending A Dime

Scott Harkins CSP, CPCU
CEO of Harkins Consulting Group, Inc.
and
"Doctor Frank" McCluskey
Dean of Online Learning
Mercy College, New York

Writers Advantage
New York   Lincoln   Shanghai

Burning for Success
How Volunteer Fire Departments Motivate Teams, Coach Leaders and
Deliver Killer Customer Service Without Spending A Dime

Writers Advantage
an imprint of iUniverse, Inc.

For information address:
iUniverse
2021 Pine Lake Road, Suite 100
Lincoln, NE 68512
www.iuniverse.com

ISBN: 0-595-24012-7

Printed in the United States of America

Contents

This book is dedicated to the hard-working men and women who volunteer their time to make a difference in their communities.

Acknowledgements

In our research for this book, we have talked to managers from different industries. We would like to acknowledge our debt to the following businesses for insights incorporated into this book: Pepsi, for the work on their mission statement; IBM, for their expertise in succession management; Fuji Film USA, for their marketing planning; Harley Davidson Corporation, for their articulation of branding ideas; Cooper Tire and Rubber, for their vision in their work on philosophy and beliefs; GE, for their work on corporate education; Dell Computer, for their integration of technology and work processes; Lens Crafters, for their understanding of customer relationship management; Amazon.com, for their work on customized marketing; Proctor and Gamble, for their work in streamlining organizational structures; Marriott Corporation, for their work on employee retention; and Case New Holland, for their work in acquisition management.

We would like to thank the American Management Association, for whom we have worked as trainers and consultants, for the opportunity to see some of the greatest American businesses from the inside. We would like to thank the members of the Graduate Business Division at Mercy College from the programs of Banking, Organizational Leadership, MBA, and Human Relationship Management for the guidance they have provided on the insights in this book. We especially thank Dr. Thomas Milton, head of the Division of Business, Dr. Peg Cucinell, Graduate Dean, Joe Girvin, and Dr. George Pawliczko, President of the American Institute of Banking, just off Wall Street.

We would like to express our gratitude to the members of the Mahopac Falls Volunteer Fire Department for showing us what can be accomplished when a group of dedicated people move in the same direction. Thanks to Marguerite and Michelle for their patience and support.

FOR TWO HUNDRED YEARS THE VOLUNTEER FIRE SER-
VICE HAS MOTIVATED TEAMS, DEVELOPED LEADERS AND
MANAGED QUALITY, ALL WITHOUT SPENDING A DIME.

THIS BOOK UNCOVERS THE TEN SECRETS OF HOW THEY
DO THIS AND SHOWS YOU HOW TO APPLY THESE PRINCI-
PLES TO YOUR BUSINESS.

IF YOU HAVE EVER LOOKED FOR "BEST PRACTICES" FOR
MOTIVATING EMPLOYEES AND INSURING CUSTOMER SAT-
ISFACTION, LOOK NO FURTHER.

WE WILL SHOW YOU HOW TO LIGHT A FIRE UNDER YOUR
ORGANIZATION AND HAVE IT BURNING FOR SUCCESS.

Burning for Success

Would you like to revitalize your business, accelerate your projects, energize your employees, and deliver killer customer service—and do all this without spending a dime? If you would, this book is for you. We will show you a model where this is done every day. We will show you how you can put the "fire" back in your workplace. For hundreds of years the volunteer fire service has provided world-class energy and enthusiasm without paying its "employees." Without financial reward and often without recognition, it has always had members who go above and beyond what is required of them, again and again. Without the ability to hire or fire, strict discipline and respect is maintained even in the most dangerous and chaotic situations. Members have such confidence in their leaders that they follow them even when doing so could result in injury or death. Would you like your firm or your team to emulate these behaviors? Would you like one-half of this energy and commitment in your workplace? Keep reading. This book will show you how to get it.

The business environment is changing rapidly and dynamically almost every day. Those that can't or don't keep pace with changes are gone. There are battles going on to find and retain the talent to manage and compete in today's business world. Loyalty seems to be a thing of the past.

The information revolution has increased the speed of most transactions. Customers expect immediate turnaround. Deals that once took months are sometimes now concluded in days. When you left the office thirty years ago, you left your business behind. There were no emails, no faxes, and no answering machines. When you went home you were finished working. When you cleared your desk and your inbox on Friday it was still that way when you got back on Monday morning. Now business is electronic, and we are operating in a global market where someone is awake somewhere and doing business twenty-four hours a day. If your email inbox is empty on Friday night, it could be full by Monday morning. The pace has picked up and the speed of doing business as usual has increased.

This increased speed has made dinosaurs out of many businesses. The virtues of business years ago were caution, thoughtfulness, and not making costly mistakes. Managers rose to the top because they made few mistakes and did not take unnecessary risks. The "yes men" (and they were mostly men in those days) who climbed up the corporate ladder by not rocking the boat were the managers to be emulated.

The information age changed all that. It was not just about getting to market but getting to market first. It was not just about getting back to your customers but getting back to them in a rapid fashion. The expression "customer relations management" really means never disappointing a customer, no matter what day they call or what time of day they call. You need to be in motion 24/7.

Once your business understands that speed is important, all the traditional, timeworn values have to change. You have to rethink how to do things, how to look at customers, and how to move at breakneck speed without making mistakes. You require new models and new paradigms. This book provides one such a model. This model is over two hundred years old. Its value for business should have been obvious decades ago. However, it will be laid out for the first time in this revolutionary book. We will show you how to move at the speed of light in the information age, without getting burned.

How can leaders and managers succeed in this new environment? Our model has a proven track record of success in an ever-changing world. It is a model where employees are highly motivated and operate on a daily basis to support the vision of the organization. It is a place where leaders, managers, and employees make split-second decisions, with life and death consequences. It is a model where employee teams take such pride in their work that they become the best advertisement for finding, retaining, and motivating members.

Where on earth can you find this model, you might ask? Look no further than the volunteer fire service. The volunteer fire service has been providing emergency services for hundreds of years. The vast majority of communities in the United States of America are receiving fire protection and emergency medical services from unpaid professionals. These organizations are able to attract and retain highly motivated "employees," excel at creating a strong sense of identity, and develop leaders who inspire trust and are able to make split second decisions in a constantly changing, chaotic environment where there is no margin for error. A volunteer fire department must do this all without the ability to financially reward its "employees." At the same time they create and maintain an environment of excellence that can be a model for *your* business.

This book will give you an insider's view of how the secrets of the volunteer fire and emergency services can be used to improve your business' bottom line. Each chapter of this book deals with an issue that all organizations face. Each chapter is self-contained; we have made it so that they can be read in any order. We suggest that you take a look at the Table of Contents to pick an issue you currently face. Read the chapter to find out how lessons learned in volunteer emergency service organizations can help you in solving problems and improving your bottom line. We're sure that once you have sampled a chapter you will want to go on to another and another. Once you get a taste of the message being delivered, you will see the benefits to you and your organization of applying lessons learned in volunteer emergency services to your organization and your life.

In this book we have traced out the **10 secrets of how the volunteer fire service produces killer customer service without spending a dime.** We are sure that you will find these ideas useful in your business.

To find out more, or to arrange a workshop, speaker, or consultation, contact:

drfrank@burningforsuccess.com
chiefscott@burningforsuccess.com
burningforsuccess.com
Or call toll free 866-347 3362

"Every success is built on the ability to do better than good enough."
—Unknown Author

"Success is peace of mind knowing you did your best."
—John Wooden

There Is No Margin For Error

"The victory of success is half won when one gains the habit of setting goals and achieving them. Even the most tedious chore will become endurable as you parade through each day convinced that every task, no matter how menial or boring, brings you closer to fulfilling your dreams."

—Og Mandino

On a scale of one to ten, rate the accuracy of the following statements, as they pertain to your organization.

- Customers stay with us because they know they will not get the same service from our competitors.
- Employees have the authority (and are willing to use it) to make decisions affecting our bottom line.
- Daily tasks are carried out with 100% attention to detail.

In business there is no margin for error. A simple computer glitch can kill a website at a crucial moment. A bid mailed a day too late can result in lost wages or even jobs. Inattention to market conditions can lead to overexpansion at the wrong time. Business is moving at the speed of light (or close to it) and key decisions must be made immediately. There is no longer time to gather all the pertinent information, evaluate all the pros and cons, and come to a consensus on a course of action. The old saying,

"He who hesitates is lost," has never been more true. The pace of business is the direct result of the revolution in information technology. As the speed of information has increased, the need for quicker decisions has increased, as well. This has changed the basic values of management. Thirty years ago the values we touted were steadiness, reliability, caution, and not rocking the boat. As we said earlier, these were the values of the "corporation man" of the 1950s. With the increase in speed came a need for different values. The businessperson of today is very different. We now praise nimbleness, the ability to turn on a dime, and the ability to foresee economic changes. A more diverse and faster moving manager has replaced the manager of the fifties. The skills we need today—fast paced decision-making, unforgiving accuracy of judgement and the ability to handle the pressures of rapidly changing situations—have a precedent. We would like to suggest a model that has successfully navigated the waters of quick decisions for generations: the volunteer fire department. *Fast Company Magazine*, one of the heralds of the new revolution in business, has a yearly article entitled, "Who is Fast?" It highlights people that are ahead of the curve and are acting in a way that will bring about the future that faces all of us. Volunteer fire departments have been "fast" for generations. They must combine speed and care in a way that will focus every decision on accomplishing the task at hand. Like your business, there is no margin for error in the volunteer fire service. The fire service is crystal clear about this. They know that any mistake can be a costly one. They know that any delay is unacceptable. They know what business they are in and what the stakes are. This gives them energy, focus, and a devotion to excellence that we can learn from.

Volunteer fire departments have faced the problems of motivating members while maintaining quality for more than two hundred years. In the world of the volunteer fire service, simple decisions can lead to injury or death. These organizations must operate at the highest level of efficiency every single minute of every day. For leaders in this world and the people who trust and depend on them, there is literally no margin for error. But even with all of these extraordinary dangers, they have additional challenges that businesses do not face. They cannot hire or fire people. They cannot give raises or take away pay. Promotions bring more work and responsibility, but little else. There is no financial reward for the "employees" (members of the organization). At the same time, these organizations motivate teams, develop and create leaders, insure excellence, and function at a very high level. Without monetary rewards, how do they do it? Without the ability to hire and fire, how do they get the job done? How do they perform the extraordinary work they do without all of the rewards, perks, and structures businesses use to do the same thing. If you stopped paying your employees today and told them there was no pension and no chance of advancement, would they be at their desk tomorrow? In most cases the answer would be no. Many people go to work primarily to earn a living. However, volunteer fire departments get ordinary people to do extraordinary things without any of these rewards. If you would like your employees to walk through fire to get the job done, keep reading.

The priorities of any organization are most clearly revealed in what the leaders *do*, not what they *say*. It is how a leader acts that will show the employees what the real priorities are. We often hear people say that "teamwork" is a priority but management then makes decisions autocratically with little or no feedback from anyone. In that kind of organization, the "talk" about teamwork may be important, but teamwork clearly is not. If your company has workshops on teambuilding and at the same time makes decisions that impact employees without letting them take part in the process, you are sending them a mixed message. Have you or senior managers in your organization ever talked about being a "world class organization" but not provide the budget, training, or personnel to reach that goal? If this is the case, the talk about a world-class organization remains only talk and cannot be realized. This is sometimes called "talking the talk without walking the walk." When goals are set, management must act in concert with these goals. After goals are set they must be acted on.

There must be buy-in on all levels. Once the message is given, management must show through their actions that this message is serious.

In the fire service it is often clear what the message is and what the consequences will be if the message is not communicated effectively. If a one hundred-foot aerial ladder is not properly maintained and two or three people must go the top of it in darkness or bad weather, the consequences can be deadly. If an air pack, which is the firefighter's lifeline to good air inside of a fire, is not maintained, the result can be death by suffocation. If a flashlight battery dies deep inside a dark burning building, it can result in a very serious situation. In the fire service you have to "walk the talk" to make sure that everyone gets to go home at the end of a call or shift.

In checking every piece of equipment and in training each member, there is a clear understanding about the importance of quality and precision. There is an absolute attention to detail in performing routine tasks, and no short cuts or compromises can be tolerated. Even small miscalculations can result in injury or death. There may not be life or death consequences in all of your tasks, but 100% attention to detail will lead to increased customer satisfaction, a safer workplace for employees, and a sense of pride in the work being completed.

You might be saying to yourself at this point, "Sure, that sounds great, and I agree with it in theory, but who has the time to pay attention to 100% of the details?" Well, if you don't have the time to do the job right the first time, how are you going to find the time to do it the second time? Firefighting is one of those worlds where every detail is the most important detail. Airline mechanics and anesthesiologists must feel the same way. Checking and double-checking is a way of life for them. The job of leadership in the fire service is to communicate this attitude of attention to details to the membership. Any leader in any business must also effectively communicate this to his or her employees. All leaders must spend time getting buy-in on what is important for their organization. If the internal customers are not on board, you can bet the external customers will notice.

The difference between firefighting and most businesses is that mistakes sometimes end up on the front page of the newspaper. That is why management and leadership theories and ideas have always been important to fire service leaders. Lessons learned in the fire service are critical and must be passed on to all leaders and to all levels of the organization. Ignoring these lessons can result in more injuries and deaths.

To you, your business is as important as firefighting is to a fire chief. You need to communicate this to your employees. They must understand how every task ties into the larger task of making the organization successful. If you do this, you will have forged an organization where the employees believe that what they do matters. Employees who feel this way will agree that there is no margin for error.

Ask yourself some questions about your business:

- Are there acceptable margins for error in my business?
- What are they and why do we accept them?
- Would my business be more successful if these errors were eliminated?
- What must be done to eliminate them? (Hint: look at the other chapters in this book. Keep notes in this space to help you remember the ideas you have found to make your business more successful.)

The next chapters will give you the 10 secrets of success that allow the volunteer fire service to do what they do 24 hours a day, 7 days a week, with enthusiasm and pride, all without spending a dime.

Secret Number 1:

Attack The Fire Without Hesitation!

"Vision without action is a daydream. Action without vision is a nightmare."

—*Japanese Proverb*

On a scale of one to ten, rate the accuracy of the following statements, as they pertain to your organization.

- Top management has created a vision that makes it clear what we stand for.
- Employees understand and embrace our vision.
- We have taken the steps necessary to operationalize our vision.

Can you clearly state the goals of your business? Is it going where you want it to go? Do your employees know these goals and what it takes to achieve them? If you answered "yes!" to all of these questions, congratulations! You and your company have a clear vision! If you answered "no" to any of these questions, then perhaps you should look at the vision you have for your company, division, department, or team.

Volunteer fire departments know what business they are in and why what they do matters. Over the hundreds of years that they have evolved into the modern organizations they are, they have learned quickly from their mistakes. They have had to learn quickly because of the potentially deadly consequences of not learning. Their first secret of success is that they have a vision, and that vision determines all of their other efforts. They know their job is to attack the fire without hesitation. All of their vision statements and all of their mission statements allow them to do just that. When the whistle blows they will respond quickly and safely with trained personnel and equipment that has been carefully maintained. When the whistle blows they don't need to be told what their job is. They have a vision and they know what has to be done to realize it. Think about this model and ask yourself what it would take to enable your employees to have the same clarity about their jobs, and how doing that job contributes to the goals of your organization.

There are two sets of questions you should be asking your employees. The first set is about their understanding of your vision. These questions include the following: Do they know the vision? Is it clear to them? Do they understand what it takes to get there? There are a number of ways to help make sure they have this understanding. Some businesses put their vision statement on the back of employee ID cards. Others plaster it in every elevator. Others have it on the back of their business cards.

The second set of questions is just as important. Are we doing what it takes to operationalize this vision? Are our actions each and every day in concert with this vision? Are we doing what needs to be done to provide direction to the organization?

Your business needs to have a clear vision. Employees need to know that what they do every single day is important. If they don't feel they are doing something that makes a difference, they will certainly look elsewhere, even if you are paying them well. People are often shocked to find out that money is not the number one motivator for job satisfaction. If someone feels what he or she is doing matters, and that he or she is being recognized for that effort, that is much more valuable than monetary rewards. Remember, as we already stated in the first chapter, volunteer fire departments do not pay their members a dime but often get them to do things that they would not do at home or at work. They do not do these things for money, for a pension, or for any other tangible reward.

It doesn't matter if you are in the business of saving lives or providing industrial clients with cleaning supplies, employees need to have a sense of self worth. This means that your employees must see that what they do on their job every minute contributes to the overall goals of the company as expressed in the vision statement. Develop and reflect a vision for your business that will allow employees to develop this sense. Your vision should allow employees to see that their job is important. There is a reason you are in business. What is it? First make sure that YOU know. Then communicate that idea to the other employees. Make sure your employees understand why they do what they do, and how that impacts your customers and the success of your business. As an example, look at BASF's advertising campaign. "We don't make any of the products…we make them better." Incorporating that into their vision will allow employees to develop a sense of belonging to an organization that is trying to make a difference. Couple that with actions by senior management to bring that vision to life and you start to create an atmosphere that will be a major factor in attracting and retaining good employees. It can make your business a place where employees want to work!

A similar story can be found at Pepsi. The company was once a very successful competitor of Coca-Cola. Trying to expand its market, Pepsi purchased a number of side businesses, including Pizza Hut, Kentucky Fried Chicken, and Burger King. The company became complex,[1] and some people said that the organization lost its focus. Some years ago the leadership at Pepsi redid their vision and asked all of their employees to take part in the process. They made illustrated organizational maps and asked for a lot of input. At one point they came up with a three-word vision statement that would get them back on track. They got rid of Pizza Hut and the other businesses they had acquired and went back to doing what they did best. Pepsi now had a new vision "WE SELL SODA." Everything in the company could now be focused on the essential business that got them to the top in the first place. They had found their voice again.

We talked earlier about the pace of change in business today. It is important to keep in mind that because of change a vision statement needs to be a living document. It profits a firm to constantly review and re-evaluate their vision and make adjustments when needed. After some acquisitions in the late 1990s, Cooper Tire and Rubber Company saw a need to reaffirm their direction. Their vision (purpose for being) for the new millennium was emphasized and published in their philosophy and beliefs.

Similarly, our fire department redid our vision some years ago. We wanted to write a vision statement that would capture the imagination of our members and focus our efforts. The first thing we did was to come to the understanding that we needed buy-in. If a single individual or small group of managers or board members concoct a vision, they are the only ones who own it. For all parties to own it, see it, and be enthusiastically onboard, the process must include input from everyone from the start.

At our firehouse we spent a long time talking about what should be included in our vision statement. What should be excluded? What should we emphasize? All of the members of our fire department were asked to take part. We included the ex-chiefs and older members to take advantage of their feedback. Months and months were spent debating, arguing, and being sure we were clear about both the language and the goals we wanted to reach. After a long process and a good deal of discussion we came to the following:

Our fire department's vision is:

"To be recognized as a leading provider of integrated emergency services."

The following statements support it:

To be recognized as a leader, the Mahopac Falls Volunteer Fire Department is committed to providing professional-level emergency services to the community in which we serve. Superior service to our community will be achieved by integrating the services we provide into the fabric of our membership.

To provide this level of emergency services, the Mahopac Falls Volunteer Fire Department must capitalize on its' greatest asset: THE MEMBERS. Excellence can only be achieved through the combined efforts of individual members with a desire for continuous improvement. We will improve by taking advantage of the memberships current skills, developing needed skills, and increasing the current base of knowledge.

To achieve this vision the membership needs to be committed to improving their knowledge of emergency services, refining their current skills, and developing the needed skills to excel in at least one aspect of our operation. In addition to their own self-improvement, the officers will provide opportunities for the membership to refine and improve needed skills.

You may have noticed some interesting aspects about our vision statement. It is short—just one sentence. You could put that on a card and keep it in your wallet. You could put it on top of your stationary. You could repeat it everyday. Second, you may have noticed that the word "volunteer" does not find a very important place in the rationale supporting the statement. That is because we want our external customers, the public we serve, to see us as professional and competent. This is our goal, and we wanted a vision statement that would focus our aim.

The vision statement is posted just inside the main entrance to our firehouse. It serves as a constant reminder to all who enter. It is a focal point of the training sessions we hold and a topic at staff meetings.

Once we adopted our vision statement it was time to get people in tune with it. This is an ongoing process. When we set up our yearly training, we want to make sure that every drill, every lesson, contributes to fulfilling our vision. This is part of the process. The second part is to make sure the members are on board and to make sure they are acting with a common purpose. It is difficult to get large groups of people with diverse backgrounds to develop a common sense of purpose. Fire departments are very successful in doing just that. Most fire departments operate with companies. (A fire service company is the same as a team in most corporations.) Internal competition is often used to foster a sense of belonging and purpose for the company members. Rewards may be offered (for the cleanest truck, most efficient use of hoses or ladders, etc.) but the key to promoting excellence is the desire of one team to be better than another team. This is pride. Pride in doing the best you can. Pride in your team for doing what you do better than anyone else.

Success in the fire service comes when the teams (companies) put their skills working as team members together to have a positive outcome at an emergency scene. How do you build for success in your organization?

Have you painted a big picture in your organization? Do your employees know what you expect of them and why that is important in making the organization successful? Expectations can make or break an organization. We generally get what we expect. Therefore, our expectations must harmonize with our vision. In business, as well as the fire service, these are critical steps to long-term success. A part of the vision in the fire service, as it should be in every business, is that everyone gets to go home at the end of the shift or tour in the same condition that they arrived for duty. The other part is providing a service or product that helps other people. Deep down we all get satisfaction from helping others.

Keeping an eye on the mission helps keep employees in the fire service satisfied. They will spend countless hours cleaning and maintaining equipment, training, and developing the skills needed to be successful, all because they are focused on what needs to be done. In our department, we have a vision statement that paints the picture for our members.

You may have a vision statement, but if you walked up to your average employee and asked them what it is and what it means to them, would they know? You may need to rethink your vision statement. In many businesses, it is not just the business that needs a vision statement but each business unit or team that needs to focus on a task or area of expertise. Fire departments know what they are about. Walk into a volunteer fire department and ask any of the members, *any of the members*, why they joined and what their job is. Ninety-nine times out of one hundred they will know and be articulate about why they are there and what their role is. They may not be the best member, or the most reliable, but they know the vision and what can happen if they don't fulfill their responsibilities. If the whistle blows, they know what their job is. There is no question about it. There is no confusion about where the organization is headed or what direction needs to be taken. They know what business they are in. How can you apply this to your business?

To help write a vision statement, think about answers to these questions/statements.

- What does your business/company stand for? What do you want it to stand for?
- What do you want your business to be known for?

- What are your organizations core values? Does top management "show" these values?
- Employees in our organization know they are important because they....

Chiefs Should Lead, Not Manage!

Leadership does not always wear the harness of compromise.
—*Woodrow Wilson*

On a scale of one to ten, rate the accuracy of the following statements, as they pertain to your organization.

- Employees trust the leaders in our organization enough to allow them to make split-second decisions that could mean the difference between life and death.
- Our organization's leaders, managers and team leaders make quick, accurate decisions.
- Our organization's leaders do not postpone decisions until all the facts are clear and the best course of action is self-evident.

When a house is on fire there is a great deal of excitement. Firefighters know that the heat in a modern fire can melt their helmets, melt their masks, and burn the coats off their backs. They go in for one reason and one reason alone. They trust their leaders. They trust them with their lives. Can you say the same thing about your leadership team? Do your employees feel loyalty and trust towards the leaders in your company?

Along with the traditional leadership, many organizations have an "underground organization." This is where the negative energy and criticism can be heard down at the coffee pot or water cooler. This is where every decision you make and even your motives are called into question every day. If this underground movement gets enough momentum, it can destroy or cripple the effectiveness of your business. This underground force has its leaders just as your visible organization does. There is a hierarchy of people who do the complaining. It has an organizational chart that is just as defined as the one you use for your organization. Somewhere, in most companies, there is a CEO of complaining. He or she may have a CIO and CFO of complaining that "report" to them and feed the flame. How can we use the leaders in this "organization" to help us promote what we do rather than destroy it?

Do you know who your leaders are? Some people in your organization are thought of as leaders based on their position (president, chief operating officer, etc.). While their role gives them "managerial authority," they must inspire the trust and confidence of the employees if they are to be effective leaders.

Leaders need to know who they are and where they are going. Leadership is action—not a position or rank—so developing leadership must start from within the person. This is the difference between a leader and a manager. Managers get things done by virtue of their position. Leaders run things by the nature of their character and the trust other people have in them. You can be an incompetent manager and still have employees reporting to you. They have to because that is their job. But you cannot lead if no one will follow you. What we look for today are people who have both leadership and managerial skills.

In the fire service we are often called upon to make decisions that could injure or even kill our members. At the same time, the decision making process must be so rapid and smooth that there is no time for discussion or debate. In these stressful situations, an officer gives an order and the members carry it out. It is that simple. Volunteer fire departments often call their decision-making process "paramilitary," meaning it is a hierarchy where there is little room for debate between those giving the orders and those carrying them out. The only time a member can disobey an order is when that member believes that the action is unsafe and may cause unnecessary risk.

To operate in such a structure there must be trust between the membership and the officers. This is where the lessons of leadership have been honed to perfection in the volunteer fire service. If an officer is not a leader, it will be evident when the officer is a new lieutenant, the lowest rank of officer in most fire departments. Because most volunteer fire departments advance their officers through an election process once a year where all of the members take part, those who do not show leadership abilities are culled out of the process early. By the time the lieutenant moves up to captain, there is more trust in the individual by the membership. By the time an officer makes it to assistant chief, there is another vote of confidence. To be chief in the fire department is a very weighty responsibility. Your words or actions could mean the lives of your members. Leadership here means not just keeping the status quo, but the ability to lead people through situations when every muscle in their body is telling them to flee.

How effective a leader is must be based on the results he/she achieves. There are many ways to become an effective leader. As we mentioned in the last chapter, to be effective, leaders need to have a clear vision for the organization. They must communicate this vision to the organization, and put life in it through their actions. Employees, just like members of a fire department, are very good at recognizing the truth. They know when someone is feeding them a line. They will certainly be more committed to and will trust the people that "give them the real deal." A leader can show that he or she is giving them the real deal by actively pursuing the vision he/she has communicated. This increased trust and commitment motivates employees to work for the common good. This means telling the truth. Most organizations have a level of dishonesty that inhibits free and honest communication. A leader is a person who can be counted on to communicate honestly and look at things as they really are. Integrity is an extremely important trait in a leader. Nothing destroys trust faster than dishonesty or a lack of integrity. Integrity is a simple thing. It means you act to achieve the goals you talk about. When you say you are going to do a task, you do it. If you make a promise, do everything in your power to keep it. Something as small as not returning a phone call when you promised to can hurt your credibility. Being some place at an appointed time can maintain the perception of your integrity. While other cultures may not be as strict about keeping appointments as North Americans, it is an important perception of people in our culture.

One reason why people may lose the trust of others is that they make promises they cannot fulfill. Unable to say "no" they make promises that are not delivered on time. Over-promising and under-delivering is the worst thing you can do. It is much better to under-promise and over-deliver. This way you exceed the expectations of those you work with. This kind of behavior builds trust. It accomplishes two purposes—it manages expectations and it gives you a track record of keeping your promises. In areas like marketing, sales tracking, and information technology, it is important to manage the expectations of your internal customers. Management sometimes wants too much, too fast. If you agree to this, you could be the one holding the bag when it does not come about. Everyone would like double-digit growth, but this is not the norm in any business all the time.

Leaders have to provide the risk-taking and entrepreneurial imagination for a company to be effective in today's rapidly changing business environment. Effective solutions require imagination. They also require leaders that have confidence in themselves and their organization. This begins with knowing they are doing the right things. They must develop a comfort level to make decisions (quickly and accurately) without all possible information being analyzed. At the pace of change in today's business world, you don't have the luxury of waiting until all possible options and consequences can be weighed. As suggested earlier, the speed of information technology has forced many of the changes in management theory. Business is becoming more like the fire and emergency services.

To have a successful outcome, you need to have leaders that can make quick, accurate decisions based on the facts that are present at any given time. For instance, every fire inside a building requires the first arriving officer (or any leader) to make a rapid size-up of the fire scene and determine whether the fire will be attacked from the inside, or whether all personnel will be positioned outside to try to contain the fire. This decision must be made rapidly and be based on the conditions present at that time (extent of the fire, life hazard to occupants, water supply availability, building construction, etc.). We don't know of any study that has looked at this, but we would venture to say that given an hour, a day, a week, or a month to gather additional information and analyze possible outcomes would not change that initial decision in very many cases. This decision may cost people their lives, yet it is still made and communicated with conviction. If the fire service can make split-second decisions to risk lives in order to save lives and property, why does it take most organizations so long to make "simple" decisions?

One consolation that should be kept in mind is the fact that leaders don't have to make the right decision all the time; they only have to make the end result come out right. That means that we always have the luxury of making mid-course corrections if they are needed. A fire officer who has committed personnel to an interior fire attack can evacuate them and take a defensive posture should conditions deteriorate. That mid-course correction could be instrumental in protecting the lives of firefighters. This should help us remember that we need to monitor the actions that result from a decision to make sure we are heading in the right direction.

Every organization has leaders that don't hold a title. What makes these people leaders? The answer is simple: their actions. They take action and others follow. This may be based on trust, past performance, knowledge, or respect. It is important to identify these people in your organization and to make use of their leadership. Let's look at an example to clarify this thought. At the scene of a fire, all of our members (except the chief officers) report to a staging area. In staging, they are assigned to a team (ventilation team, hose team, ladder team, etc.). A team leader is assigned (based on the individual's past performance, knowledge, and trust) to direct and oversee the team's actions. These individual team leaders are instrumental in achieving a successful outcome. This means that the person leading the hose team into the fire is as important as the chief when he or she decides to be aggressive or hold back on the attack. No matter what the orders of the chief are, the membership may look to the leader who is an ordinary firefighter before they will feel comfortable with the decision of the chief. Leaders run fire-ground operations. They are no less important in your organization.

Here are some exercises to help make you a more effective leader:

- Identify the leaders in your organization: (don't forget the ones who don't have leader-type titles).
- What are their strengths?
- How can you make use of them to move your organization toward your vision? (this includes members of the underground organization.)

Secret Number 3:

Firefighters Trust Their Teams With Their Very Lives!

"Never doubt that a small group of thoughtful, committed people can change the world. Indeed, it is the only thing that ever has! "
                                                               —*Margaret Meade*

On a scale of one to ten, rate the accuracy of the following statements, as they pertain to your organization.

- Expectations are clear to our employees and employee teams about what must be done and by when.
- Our employees feel that they have the authority to accomplish tasks without constantly checking back or fearing they will get in trouble for making a mistake.

- There is a coherent team structure in our business, and our teams are staffed with the right people.
- Our teams have been provided with the resources needed to accomplish their goal(s).

How did you rate yourself? If you are a ten in each area, congratulations, but you are in the minority. One skill most businesses do not do as well as they could is harnessing the synergies of their people working in teams. There are many skills that make for a great fire department, but none is more important than the ability to work as a team. Fire departments depend on teamwork to be successful. No matter how talented any individual is, he or she cannot possibly put out a fire by him or herself.

Teamwork is at the heart of firefighting. Firefighters do everything together. Individuality is always bent to the will of the group. Every task involves the cooperation of members and the sublimation of their own desires so that the team can accomplish its mission. A hundred years ago, when there were no female firefighters, the fire service was called a "brotherhood." Quite simply, this meant that fire departments were experts at building and maintaining teams. The team building skills of the fire service has been refined for centuries. To illustrate what we mean, we should take an in-depth look at how teams operate in an emergency mode.

The chief pulls up first and uses a radio to alert other responding units that there is a house on fire. Then comes what we call "the first due engine." This engine pulls up to the scene with the initial attack crew, who will go inside to "battle" the fire. They must come off that engine as a team. They must be as organized and choreographed as a ballet company or a good football play. The fire engine stops at the beginning of the street, and one member jumps off to pull a supply line (a large diameter hose used to supply water from truck to truck) off the back of the truck. The engine then proceeds to the fire laying out this hose along the way. The member will walk behind the truck to make sure the hose is coming off correctly. When the truck stops, he/she will disconnect the hose from the hose bed and hook it into the side of the truck so that water can go from other arriving trucks into the engine's water tank to keep it supplied. Another team member will take the nozzle end of a hose line and "stretch" it to the fire building. A second member makes sure the hose is coming out straight and does not have any kinks in it. The team will make sure that there is enough hose pulled so that when the member reaches the front door with the nozzle, there is enough additional hose to operate with inside the building. If not pre-connected, a team member will uncouple it and hook it to the engine's pump discharge so that water can go to the fire. A third team member joins the member at the nozzle with tools to force entry to the building, if necessary. The person that pulled the hose from the back of the truck will also join the other members to assist in the initial fire attack. The team signals the engine's driver, who is now at the pump panel to send water through the hose line. The driver pulls the levers that will allow water to flow from his tank through the hose, and the attack is ready to begin. This can all be done in about as much time as it took you to read this paragraph.

This is only part of what is happening. Burning buildings are often extremely hot and are filled with toxic gases. They are therefore not safe to enter without "venting" the fire, or creating an opening so that the heat and smoke will escape. Fire builds up heat and smoke inside any closed structure. If a hole can be created where the smoke and heat can exit then the firefighters can have an easier time getting to the fire and putting it out. Venting, by breaking a window or cutting a hole in the roof away from our point of attack, helps make the building safer for the initial attack team. When we vent, the fire will move in the direction of the ventilation hole. This helps move the heat and smoke in that direction and makes the area where we are going to enter the building more tenable. One potential problem with ventilating a building is that it will allow the fire to burn more intensely. This is because we have now made sure there is an ample supply of oxygen that will feed the fire and increase its power.

Regardless of whether we are breaking a window or sending a team to the roof to cut a ventilation hole, we know that it must be timed with the attack. Venting too soon will allow the fire to grow, making it more difficult to extinguish. Venting too late will put the firefighters entering the building in jeopardy, as the rush of oxygen to the fire from the door or window they enter could cause a backdraft explosion. Everything must be a tightly choreographed effort to make it work.

At the same time, other teams are performing necessary tasks. All of these tasks are aimed at putting out the fire and some essential tasks may take place miles away from the fire. In our rural hamlet there are few fire hydrants, so we must shuttle water on large trucks called "tankers." A source truck goes to a pond or lake and begins to suck water. Tankers then line up to have their water tanks filled. They will rotate to the fire scene where they take turns supplying water to the attack engine. When they are empty they will return to the water source to be refilled. The number of trucks needed at a large fire can create a complex traffic flow. Everything must be coordinated correctly to keep water flowing onto the fire. In order to manage this "tanker shuttle," someone must know which trucks are full, which are in route to be filled, and how quickly the water is being used at the fire. This is called "water management."

At every emergency scene there are other teams at work. If it is dark, the scene must be lighted. The ambulance must be there to provide first aid and rehabilitation to the other teams. A safety officer (or team if it's a large scene) oversees the various operations. Teams are assigned to perform overhaul and salvage to protect property and make sure there are no "hot spots" that could cause the fire to rekindle. The air bottles we use to breathe with must be changed and refilled. There may be twenty to thirty jobs being completed by seven or eight teams at any given time, all of them vital to our success.

How does our fire department build teamwork? First, we have a very clear command structure. There is a chief, assistant chiefs, captains, lieutenants, and so on. Even our specialty teams (EMS, fire police, and dive) have their own captains and lieutenants. These are the people that we look to as leaders for the teams we form at an incident. The highest-ranking officer will be the Incident Commander (IC). If none are present, past chiefs or senior members will fill the role. At our emergency scenes there is a definite hierarchy. Only one person can be in command. That way we are all following the same game plan for that incident. On the orders from the IC, other teams will be formed to handle specific tasks. Another officer or senior member will head up these teams.

What makes for successful teamwork? The team has a specific task it must accomplish in a specific time. Several things have to happen for the team to accomplish that task. First, the team must be educated about what it is they are doing and how to do it (i.e., they need to have the knowledge and skill set to complete the task). Second, the team must have specific instructions on what things need to be done (i.e., they need to know what is expected of them). This means that the Incident Commander running the scene must give specific instructions on what he or she wants and expects. Saying, "get the hose" may not produce the same result as saying, "I want 150 feet of the pre-connected hose brought to the side door with the forcible entry tools." That second sentence tells the team what the expectation is in much more detail. Third, the team must be provided with the right tools (resources) to do the job.

Making your expectations clear is the most important thing you can do at any emergency situation. These expectations should be made clear long before there is an emergency. The team must know that excellence is required and that mistakes can be costly. They must understand that this is not a game. At the scene itself, clear communication is essential. It is important for the person in charge to communicate exactly what they want done and in what way. In our department we use "echoing" of radio transmissions to be sure the communication is clear. An order will be given, and the person receiving the order will feedback (echo) what he/she heard to assure the message was received as intended. Echoing is to emergency services what paraphrasing is to listening skills. By making use of echoing, being specific and exact, we set expectations that are not lost in the confusion of communication that is an emergency scene. In the heat of a disaster it is important that we clearly and calmly lay out exactly what we need to have done.

We operate under an Incident Command System (ICS). No one does anything alone. Teams are formed and every team has a leader who manages that group. That person has the authority to make decisions as long as he or she communicates them to the Incident Commander. The IC needs that data to stay on top of the situation. The Incident Commander cannot micromanage. He or she must trust the team leaders and teams once the specific instructions have been given. The Incident Commander's job is to oversee, not do, so trust is the key element. Do you have managers or leaders that micromanage? Do you trust your teams to get the job done correctly, on time? In the fire service we have found that the more the chief micromanages the less efficiently our teams operate. Our greatest chiefs have been those who recognize their job is to plan and oversee, not grab a nozzle or an axe. By doing this we build trust and confidence, not just in the team, but also, more importantly, in the leader.

Teamwork needs to include a clear understanding of what the team will do, how it will do it, how long it should take, why it is being done, and what happens when the task is completed. For our teams to be successful, the right people have to be on the teams. All of our incidents have a staging area where we assign people to tasks. The Staging Officer, who sets up the teams, knows that some people work better on roofs. Some people like to be inside fires. Some people don't like too much blood. At the command of the IC, our Staging Officer puts together the right teams to make sure all of the necessary tasks get completed correctly, on time. Once completed, the team will report back to Staging or Rehabilitation (an area staffed by an emergency medical team to evaluate the condition of the team and determine if they can be assigned other tasks).

The fire department operates with split-second timing in life-and-death situations. The key to our success is the ability of our teams to function independently and in conjunction with other teams to accomplish a common goal. They are put together to accomplish individual short-term goals that, when combined with others' work, bring us to a successful conclusion of an incident. Providing teams with the necessary resources and direction, allowing them to use their own ingenuity, and interacting with others while keeping the Incident Commander informed is what allows us to do extraordinary things with a group of "ordinary" people.

Here are some exercises to help you make better use of your teams:

- Are your teams performing at the level you expect? If not, identify three things that are stumbling blocks to team performance in your organization:
- Identify ways to eliminate those stumbling blocks:
- Identify five things you could do to unleash the power of your teams:

There Are No Small Details At An Emergency Scene!

"The greatest sin is inattention."

—*Zen saying.*

On a scale of one to ten, rate the accuracy of the following statements, as they pertain to your organization.

- Our company has minimized our equipment downtime and re-work of manufactured parts.
- Our company seldom has merchandise returned from our customers and always gets repeat business.
- Our customer service is excellent.
- Our facility is a safe place to work.

In this book we have said there is no margin for error. To be a "world class" organization you need to pay attention to details. If you are a global company you have most likely investigated International Standards Organization (ISO) certification and understand the importance of details in the certification process. Insurance companies that may be willing to share some of your risk from a financial standpoint are interested in what you are doing from a quality, safety, and risk management standpoint. Paying attention to details will make you more attractive to insurance companies and can help you get the best pricing for risks you would like them to share with you. In investigating accidents or conditions that exist in your facility, details can help you in preventing future accidents that can disrupt production and/or service.

Don't think details can be important? Let us tell you a story about how a little detail can have a big impact. We remember reading years ago about how the Watergate burglars were caught. When the burglars broke into the Democratic National Headquarters in the Watergate Hotel, they put a piece of masking tape over the lock on the door so it would not lock behind them. But instead of putting that tape on the inside of the door, they left a small piece of it sticking out on the hall side. A security guard noticed the tape. He called the Washington police. They arrested the burglars. The burglars demanded hush money from President Nixon. Nixon paid the burglars and got caught doing it. The Attorney General of the United States resigned. Congress moved to impeach the President. Nixon resigned. President Ford pardoned him. People got so angry with this that Ford lost the election to Jimmy Carter. The Iranian hostage crisis and recession followed. People felt disappointed and elected Ronald Reagan President instead. He began the Reagan revolution and appointed a conservative Supreme Court. America will never be the same again. Would any of this have happened if that burglar had paid attention to the details and been a little more careful in putting the tape on straight?

Our fire department provides emergency medical services to our community. Operating two ambulances gives members many opportunities to provide life-saving interventions. It also opens our department up to a number of liabilities. These liabilities are financial (potential for lawsuits), physical (injury to members, damage to vehicles or equipment), and emotional (critical incident stress).

Let's look at each of these areas as examples of why details are important. Newspapers and magazines regularly report on lawsuits against EMS providers. Equipment failure or failure to respond in a timely manner can result in a lawsuit. To help prevent lawsuits of this type, we pay attention to details. Documented pre-shift checks of the ambulance are carried out to make sure that oxygen cylinders are full, critical equipment is working properly, and the ambulance is stocked. Post-run checks are also completed to assure everything is ready for the next call. Often we, like the people on all jobs, forget the importance of details. We develop bad habits and tend to "blow" quickly through the post-run check. We must constantly remind ourselves of the importance of checking each item as if we were looking at it for the first time. It is an important job of all of our officers to constantly remind the members of the need for quality in every single act.

Physical injuries and damage to equipment are major concerns. Pre-shift vehicle checks are completed to assure the vehicles are in proper working order. Providing the necessary tools and equipment to do the job in a safe manner helps control physical injuries. Training (patient handling techniques, driving emergency vehicles) is used to teach new skills and reinforce existing ones.

There is an emotional toll on persons who ride ambulances and fire trucks. You see consequences that other people never even think about. The details that are needed in this area include de-briefing particularly difficult calls (children, and severe or multiple traumas) to help individuals deal with their emotions.

When residents call 911 looking for an ambulance, they are expecting someone who knows what they are doing to show up quickly. These people are our customers, and like your business, we must meet their expectations. We may not be concerned with repeat business or cross selling, but meeting our customer's expectations helps us prevent losses to the corporation's bottom line. Who would you be more likely to bring a lawsuit against: an ambulance crew that showed up at your door neat in appearance, carrying equipment with them, providing quick, courteous, compassionate care for the injured and their family, or the crew that didn't look professional and didn't seem to care about the dignity of the injured person or family? The only way to insure that your ambulance crew (or your service team, or product) is the former is to pay attention to the details.

What do your customers expect from you? (A quality product at a fair price, responsive professional service and support, etc.) Regardless of their expectations, it is imperative that you meet or exceed them. That can only be accomplished by paying attention to the details. If you want to get repeat business, want to build a positive reputation in your industry, and want to have a good reputation in the world, don't forget the details. Don't rest on your past performance, because your next sales call, product delivery, or service visit is what you will be judged on. Don't leave it to chance; pay attention to the details.

Here are some details to think about:

- Are your employees properly trained to do their job?
- Do they understand their job, how to do it, and the importance of doing it correctly?
- Are the necessary resources provided?
- Have you played "what if?" to evaluate potential exposures you face, and develop contingency plans?
- Do you have preventive maintenance programs in place to prevent unplanned shutdowns?
- Do/Does your shop, warehouse, company vehicles, etc. look professional?

- Do your people that come into contact with customers portray your company in a professional light?
- Do people calling your company find it easy to talk to a person, or are they being handled by machine? (Killer customer service isn't provided by machines!)

Somebody Has To Grab The Hoseline!

"The credit belongs to those people who are actually in the arena…who know the great enthusiasms, the great devotions to a worthy cause; who at best, know the triumph of high achievement; and who, at worst, fail while daring greatly…so that their place shall never be with those cold and timid souls who know neither victory nor defeat."

—Theodore Roosevelt

On a scale of one to ten, rate the accuracy of the following statements as they pertain to your organization:

- Our employees are focused.
- Our teams consistently meet or exceed their goals.
- Egos do not get in the way of accomplishing tasks in our company.

If you didn't rate yourself as a 10 across the board then you need to think about how to better manage your workers. You should think about ways to hold them accountable. Your teams should be accountable for the triumph, accountable for the failure, and accountable for the effort. Effort and vision are the keys. We should be rewarding excellent failures more than mediocre successes if we want to excel. That, in essence, is what we do in the fire service. We reward our teams for the effort, and hold them accountable for it. That doesn't always lead to the outcome we want (people and firefighters still die in fires), but with strong effort, 100% of the time, we have given ourselves the best chance for a successful outcome. If a team makes a great effort to put out a fire, yet fails, that should be honored more than a team that succeeds in putting out a small non-life-threatening rubbish fire. For businesses to be wildly successful, people cannot be afraid to fail.

One of the things firefighters pride themselves on is that they are people who are not afraid to take chances. When there are terrible situations that might well paralyze the average person, firefighters act. No matter what kind of management you have at a fire, nothing will happen unless somebody grabs the hose and gets things started.

In the fire service, jobs must be done in a very exact order. Let us take as an example freeing someone from a vehicle that has crashed and turned over. Many tasks are required for a successful outcome. Lights are needed to illuminate a call that happens at night, tools must be brought to the scene, the car must be kept from rocking back and forth, the car must be "opened" up, and the medical team must be able to treat and remove the patient from the car and bring him or her to the hospital. In order to have a successful outcome these tasks must be done in the right sequence, and must be done with precision and care. These tasks need to be benchmarked long before the emergency so we know how long a particular task will take and what variations we might anticipate.

The Incident Commander (IC) must know who is doing what task and what stage of completeness the task is in, in order to move ahead. Like the conductor of a symphony orchestra leading separate instrumental groups and blending them so the audience gets to hear beautiful music, the IC must know who is leading each team and how long the operations will take to get the victims to the hospital without any further injury. This is crucial for success. In the fire service, each team is held accountable for getting their job done. Like a second violin that is off key, a team that is not doing their part for the whole will make the changes needed for the team to be successful, or be replaced.

Accountability has a double meaning in the fire service. First, it means knowing where your people are so that they are always protected. Second, it means knowing who is assigned to each task so that he/she can be held responsible for its completion.

When an IC assigns a team to a particular task, that team should understand how their task fits into the whole picture of the emergency event. In our training, we reinforce how each part fits into the whole. We constantly challenge our members to understand how things that appear small or unimportant can cripple a whole lifesaving operation. If we do everything in the extrication process correctly and transport the person to the ambulance, it is all in vain if the ambulance is out of gas or the oxygen tank is empty.

Our teams also realize that the importance of their task makes it imperative that they accomplish their mission on time. It is not a question of ego. It is not a question of control. It is a question of getting the job done correctly, on time. They know that what they do can make a big difference in how safe everyone else is while doing their tasks. Do your teams and team leaders have this sense of urgency? Shouldn't they?

In the fire service, once we know what team is assigned to which task, what tools they have, and how long the job should take, we can hold the team accountable for accomplishing its task, or at the least, giving it their best efforts. Failure by any team lessens the possibility of an optimum outcome. It also increases the demands on other teams working the incident. Peer pressure, IC to team leader mentoring, and the informal evaluation of personnel by the other members of the organization all play a part in the accountability of the team. Excellence is a culture that must be constantly monitored and nurtured.

Some years ago we were operating at a fire in an area where there are no fire hydrants. A large number of tanker trucks were in a line emptying their tanks to get water to the fire. A truck pulled up and the driver/pump operator was having trouble getting the truck to pump so that he could supply the trucks at the fire. The officer (team leader) in charge of the task came up to the driver and said, "Move it along; we will use the next truck." The driver did not want to go without leaving his water. "Just one more minute." he said. The officer guided him towards the driver's seat and said, "This is not a drill. We don't have a minute." The officer understood that this was not about that driver's feelings, but about getting the job done.

Ask yourself the following questions:

- Are teams/team leaders held accountable for results they achieve? How?
- Are teams/team leaders held accountable for the efforts they put forth to achieve the desired results? How?
- Does our compensation system reward the efforts or only the end results?
- Would rewarding the efforts be more beneficial to sustained success of the organization?
- How can we modify our processes to better hold people accountable?

Secret Number 6:

Firefighters Advertise Themselves!

Pride is a personal commitment; it is an attitude that separates excellence from mediocrity. Are your employees proud to work for you?

On a scale of one to ten, rate the accuracy of the following statements, as they pertain to your organization.

- Employees at our firm feel loyalty to the firm.
- Employees at our firm go the extra mile to satisfy the customer or fix the problem.
- Employees at our firm feel they are a part of the process.

If your firm is like many others, you didn't give yourself all tens. What could you do to improve the performance of your employees? How could you make them feel more a part of the process? How could you get them to take a more active role in your business?

Years ago, the whole idea of running a business was very different. There was a boss and it was the boss's job to make sure things were done right. Responsibility was clearly defined and people knew their roles. The expression "that's not my job" made more sense in that world. But in today's business, there is an emphasis on employee empowerment and responsibility. Often times this is more a matter of rhetoric than reality. Empowerment often means you have new responsibilities but the same resources and authority. If you increase responsibility, but do not increase resources or authority, you are simply creating more stress on the work processes and the employees. Just as you want the employees to trust the company, you must trust them with the tools to get the job done. One of the rules sometimes spoken about in management seminars is "unlock your tools." This means not only that you trust the employees not to steal the tools, but that they have access to what they need if they want to self-start a project. True empowerment means you not only share the responsibility but also the power and the rewards. Unless all three are tied together you have not empowered a work team.

The speed of the information revolution has made it harder and harder to check with the boss before making a decision. When the world turned slower, it made more sense to check every decision up the line before anyone took a risk. That was the importance of the chain of command. Today's business world moves a lot faster. Customers no longer want to wait until someone checks with five other people to solve a problem. Expressions like "killer customer service," "owning the customer experience," and "total customer satisfaction" show us that the world we live in is filled with people who have higher expectations.

To meet customer expectations we need employees who are as committed to the company's success as their leadership is. They need to identify with your firm as they identify with a sports team they root for or their family. They need to feel good about belonging to your company.

For centuries, volunteer fire departments have been getting great employee loyalty without paying their members a salary. How is it that they can motivate their members to join and participate with great energy without the rewards we normally think about in business?

Volunteer fire departments have learned how to build a brand. While discussions of branding and brand-building are all the rage in modern business literature, volunteer fire departments have been doing this well for over a hundred years. Here are some of the branding strategies they follow.

If you were to walk into any small town you could easily find the people who are members of the fire department. This does not go just for volunteers. Career firefighters are often seen around our town wearing their department baseball caps, windbreakers, sweatshirts, or t-shirts. You can tell the volunteers not just by the blue or red lights on their car, but by license plates and window and bumper stickers. Everything about them is screaming that they are proud to belong to their organization. But it does not stop there. If you were to visit the homes of many of these firefighters, you would find plaques, beer mugs, coffee mugs, plates, statues, and the list goes on. When a volunteer firefighter visits another town and stops into another firehouse, they often show pride in their own company and talk about how well they are doing. Each firehouse has its own patch with a logo that members wear. These are often exchanged and traded like you would baseball cards from a favorite team or a tee shirt from a vacation spot.

Imagine your employees feeling about your firm the way volunteers feel about their fire department. Imagine the energy you could unleash if your employees felt that every one of your customers was their customer. What if your employees treated every customer as if he or she were the only customer? Try treating your own employees as if THEY were the customers. Imagine how you would feel and how well your business would do with this new energy and loyalty. Your brand is your flag. Just as some nations have made wars to carry their flag all over the world, you would like your customers, and better yet, potential customers, to know what you are about. Your brand tells them that. Fire departments have done well building brands. How have they done this? The following three things play a major role:

1. Everyone understands the mission of the department.
2. People can make a difference.
3. Members are rewarded in small ways that keep them motivated.

Let's take a closer look at each of these:

The mission must be clear to everyone. A volunteer fire department has a simple mission. When there is a call for help the fire department shows up with properly trained people and the right equipment to solve the problem. There are a number of small steps that go into this larger issue. Many small details have to be taken care of to enable this mission to be accomplished. Fire departments have a long tradition of teaching members that every small detail contributes to the larger mission. When doing something as simple as a weekly check of the tires, it is not unusual for the instructor to tell the new member that if that tire fails, the truck will not get out. If the truck does not get out, precious time could be wasted. If that truck does not respond as planned, lives and property could be lost. Now, suddenly, something as mundane and boring as a weekly check of the tires can be seen as a matter of life and death.

How have you sold the mission of your company to your employees? Have you sold them on how important it is that they do the little things right?

Do you provide ongoing training that connects each task or detail with your mission? Do you reinforce the goals that drive every activity? Do your employees see the connection between what they do every day and how the company will succeed or fail?

Fire departments believe that everyone makes a difference. Everyone is a salesman. This may not have been true in the old management style but it is vitally true today. People must be sold on the idea that they matter. Your employees must "sell" their ideas to middle and upper management. A project leader must "sell" the project's importance to the team that may have many other responsibilities. People want to feel valued. They want to feel that they are a part of a team. People want to know that what they do will be appreciated.

Volunteer fire departments expect a great deal from their members. Departments expect their members to get out of bed at all hours of the night and sometimes stay out for long periods of time in the cold or freezing rain. They ask their members to risk their lives. They ask them to give up holidays and quality time with families. In return for this, we do not give them any monetary reward. So why do people volunteer? They do it because they feel appreciated. They do it because someone says please or thank you. *Members do it because they know that the leadership of the department understands that they make a difference.*

In your business, have you done enough to show your employees that they matter? Do they know that their contribution is appreciated? Do they know that you understand that what they do really matters?

Volunteer fire departments reward members in small ways to keep them motivated. Volunteer fire departments don't give their members money, but they do give them small mementos to show them they appreciate what they do. Sometimes it is something as small as a baseball cap with a logo on it. But that logo does a number of jobs. First, it tells the member thank you. By wearing the cap, the member identifies with the larger whole that is the department. Fire departments are famous for certificates and awards. If you go to an annual fire department dinner you will see a great number of awards, plaques, and announcements appreciating all that the members do. Does your department have a logo? Do your employees identify with it? If they drank coffee out of mug that had your name on it would they think about your company differently?

You do not need to be Microsoft to have a brand. Every company should have one. It is the flag around which the employees focus and identify. It is the logo in which members see themselves and their work reflected. It is a rallying point for them to show their pride and help your company be more successful.

Firefighters identify so strongly with the company brand that they often spend their own money to buy tee shirts or hats with the department logo on it. They will buy license plates that say they are a fire department member and proudly put them on the front and back of their car. They are often proud of the identification they feel with the fire department brand. Why is this identification so strong? That logo identifies that person as someone special. It is not unusual that when two firefighters meet for the first time, they will feel a kinship with each other that transcends many other examples of mutual hobbies or interests. To say you are a volunteer firefighter identifies you as a certain kind of person—a person that cares about others, is willing to give back to the community, and would sacrifice their time and energy to help someone in need. No wonder this brand is one that people are proud to identify with!

Every company can learn a lesson here. What kind of person should work for your company? What identification would you like them to make? How would you like them to see themselves in relation to the rest of the company? There is a story told about Steven Jobs when he was chairman of Apple Computer. He wanted to break from the old tradition and make a new machine. So he chartered a team to go to work on this project and gave them carte blanche to do what they wanted to get the job done. The first thing they did was to raise the skull and crossbones, the pirate flag, over their team offices. This sent a signal. When they were done they gave the world the Macintosh and a whole new way of computing was born. People will not be afraid to step out as long as they can be reasonably sure that they are not walking the plank.

Think of what your core business is. Think of how you could brand that core business so that you would stand out from your competition. Think of selling that idea first to your employees and then to your customers. Before the customers will buy that idea, the employees must themselves buy it and be able to sell your brand.

- What is your brand?
- How recognized is it in your community/industry?
- Do your customers and employees know it?
- Identify five things you could do to build/improve your brand?

Secret Number 7:

Firefighters Know Their Mission!

I like the dreams of the future, better than the history of the past.
—Thomas Jefferson

On a scale of one to ten, rate the accuracy of the following statements, as they pertain to your organization.

- In spite of the challenges of working across team lines, reporting relationships in our organization are clear.
- Our organization is flat.
- People that work at our company are comfortable working outside their department, function, and geographical area.

There was a time not so long ago when businesses operated in hierarchies. This meant that as you went up the corporate ladder, more and more people worked for you. You gained more and more direct reports as your career advanced. You had direct power over people to hire, fire, and to whom you gave raises and promotions. This allowed managers to achieve the desired results using rewards and punishments, even if they lacked motivational and management skills. As organizations flattened out there were less and less direct reports, and more organizations became matrices. There were many reasons for this, but as businesses grew more complex a matrix was the only reasonable evolution for that business. For example, if a company merged its product lines and grew internationally at the same time, there could be questions of how to set up reporting relationships. If a soap company added a line of beauty products while it was expanding distribution channels in a dozen more countries, we have two kinds of growth that must overlap. The company cannot be divided up either by product line or geography. Both of these must overlap in an organizational matrix. Thus, the relation between the people who market the shampoo must be with both the people who manufacture the shampoo and those sales people in the field that may represent the entire product line. This means that the organization must matrix both skills and reporting.

In a matrixed organization, it is sometimes hard to remember what the mission is. But for any non-hierarchical team to work it must keep focused on the mission, and it must know why all of the tasks are being done. This team must always keep the mission in mind so that the goals of all actions are clear. All of us have seen situations where an operation was ground to a halt because of some small detail where an employee got bogged down. When keeping the mission in mind we must remember the old adage: "When you are up to your neck in alligators, it is sometimes hard to remember why you decided to drain the swamp in the first place."

Volunteer fire departments have always been a strange kind of organization. They have elements of a paramilitary organization, in which anything a senior officer says goes, and an organization where independent teams can take the initiative and make something happen while keeping their higher-ups informed. Let's take the example of an under-water rescue or dive call. There are many tasks that have to be performed for this to happen, and all of the tasks are related in a matrix. There is a hierarchy of tasks so there is no confusion about which tasks get priority in terms of resources and time. Let us see how these priorities are established so we can better understand the matrixed organization.

Dive calls are events that involve very precise co-ordination of all of the team members and functions, so that we can be sure that no time or effort is wasted. When our dive team is called it is usually because of someone in distress in the water or because someone or something has gone under the water. For example, we have been called out for the following: "car through the guard rail and into the lake and sinking at this time." From this information we do not yet know if the passengers got out or not. We do not know if the driver is conscious or not. We do not know how many people may be in the car at this time. Because we can never assume in our business, we must mobilize for a major incident in the water.

Our dive team, while housed in the fire department, has its own way of operating, and a special set of rules and procedures applies to a water rescue call. For example, we always wear our fire coats, helmets, and bunker pants to fires and jaws calls. But on the beach or on bridges we do not wear them. This is because if we fall into the water we would drown with all of that weight on us. So the divers are in their diving gear and all others must wear personal flotation devices (PFDs) so that they would float should they slip into the water.

When the call comes in, divers arrive at the firehouse. They hook up our dive trailer to the brush truck. That dive trailer contains all of the gear and equipment we will need for the dive. Each of the divers then hook up with another member called a "tender." This member will hold the rope that is used to communicate with and guide the diver when they go under the water to rescue those in distress or search for those already under the water.

The diver and tender will ride in the rescue truck to get their gear on while in route to the call. The diver, assisted by their tender, gets on his or her dive equipment. They then get hooked onto a long reel of rope that will be their lifeline.

When the rescue truck arrives, the divers are ready to get into the water. Someone trapped in a car below the surface with little air needs assistance immediately. The divers are in a rapid deployment mode. This means that one diver goes right into the water to the scene, while a second diver acts as his or her backup in case there is trouble. A third diver is kept 90% ready in case the first two divers get into trouble.

As soon as the diver gets into the water he/she do an underwater search to find the victim. This is done with the aid of the tender, who guides them with the attached rope. They search, mostly by sense of feel, foot-by-foot, while moving along the bottom.

While the divers are searching, other steps are being taken to secure the beach and set up support for a more long-term operation. A beach master is appointed to monitor who goes on and off the beach. Someone else records the times the divers go down and how much air they have in their tanks so they can calculate when to pull them out. EMS support is set up to provide assistance to any victims found. Lighting is sometimes required. All of this must be done with speed and precision.

While keeping an eye on the rescue, the IC must coordinate competing requests for resources. The prioritizing of these requests is often called "triage," from the French word that means to "sort out."

There are many complicated decisions for the IC, and a number of choices must be made. People and situations are competing for the same resources and there is constant pressure to re-prioritize tasks. At times like this it is important to keep in mind what the goal is. This helps determine who and what should get our attention first.

It is the same in today's matrixed organization. There are many areas that are competing for limited resources. There may be staff and operations, there may be headquarters and field, there may be sales and marketing. They do not report to each other but operate in a matrix where one influences the other. In order to operate successfully we must understand what has priority and what supports that effort. Corporations sometimes forget what business they are in. They sometimes forget who the customer is both internally and externally. The matrix begins to set its own priorities that can create goals or objectives that conflict with the mission statement.

Answering the following questions can help you best align that matrix:

- What business are you in? (This might not be as obvious as it seems).
- Have you identified your customers?
- What is your vision/mission statement?
- What is imperative for you achieve that mission/vision? (Identify at least three imperatives)
- How can you align your matrix to accomplish your mission?

Secret Number 8:

Every Firefighter
Must Be A Strong Link!

"'Employees are our greatest asset.' Bet you've read that one before. Well, it's not true for a public service firm. In Public Service Firm land: People are the only asset. Period."

—*Tom Peters*

On a scale of one to ten, rate the accuracy of the following statements, as they pertain to your organization.

- Our organization is able to attract committed employees.
- Our turnover ratio is low.
- Because of our ability to retain employees we are able to minimize the amount of money we spend to recruit, hire and train new employees?
- We are not having trouble keeping our "good" employees.

Employers often underestimate the cost of turnover. At a recent talk, an executive of a Fortune 100 Company was discussing how much it costs to hire and train a replacement for a top management position in his firm. After the audience had made several guesses, he told us that they did a cost analysis and found that it costs about five hundred thousand dollars! It might not be that much in your firm, but do you have a handle on what it does cost you? Think of all of the money, experience, and time that could be saved by retaining the right employees.

Volunteer fire departments are able to attract and retain committed employees. These volunteers will perform tasks for us that many of them won't even do when they are away from the firehouse. Tasks such as checking a vehicle, washing windows and floors, etc. are carried out with gusto by people who don't even pick up their own dirty laundry at home. How do fire departments do that without spending any money on salaries?

Fire departments are able to motivate their members to perform tasks in a variety of ways. First, volunteer fire departments have excelled at doing work in a way that does not make it feel like a job. Work is often done in a lighthearted way, with a cup of coffee and a doughnut in hand. The members have built trust in each other over the years by working side by side in dangerous situations, so when one picks up a shovel or a mop it is only natural to lend a hand. The volunteer fire service is fun. Someone once called it the "poor man's country club" because of the atmosphere that often exists there. For a few hours of volunteering every week, the members have a key to a place they feel is their own. They can watch TV, sit and have a cup of coffee, or just come in to visit whenever they are passing by. In addition, as we have been saying, their officers keep them focused on the big picture, create a sense of identity for the members, and allow them a sense of pride in belonging to a truly worthwhile organization. These are the areas of management and motivation in which emergency service organizations excel.

In the fire and emergency medical service business, the mission is being prepared to handle the major fire or save someone's life. Members hear about the exploits of other members who have been doing this for a longer time, and relish the possibility of being heroes themselves. The truth is…all these people are heroes, every day. They may not be the one who carries the baby out of the burning house providing mouth-to-mouth resuscitation on the way, but they have all certainly played a part in getting that baby out. The person who assured that the air pack was clean and working, the people providing fire suppression so the rescue could be made, the Incident Commander and his/her staff, and the people who made sure the tools were clean and rust-free all played a role in rescuing that baby. Understanding our mission makes all those tasks that could otherwise be considered mundane and uninteresting fulfilling. The connection between everyday boring and seemingly unimportant tasks and organizational goals must be made! It might be said that when the man in the mailroom or the woman in the copy room knows his or her task is crucial, then they understand the mission and direction of the business. It is when they do not see the connection that businesses fail or do not achieve their potential. If you walk down the aisles of any business and see people sitting in a stupor waiting for the next coffee break, the connection has not been made. Making this connection is the difference between a good company and a mediocre one. How do you do that in your company? Your business needs to have a clear vision. Employees need to know that what they do every single day is important. If they don't feel that they are doing something that makes a difference, they will certainly look elsewhere, even if you are paying them well. It doesn't matter if you are in the business of saving lives or manufacturing cars, employees need to have a sense of self worth. They must not think that they are just a cog in a wheel that does not turn very well. If they talk in terms of "us" and "them," and if there is a gulf of communication or trust between management and employees, they will not be driving your results in the right direction. No matter what you make or what kind of services you provide, you must first "sell" the idea to your own employees. If you cannot get buy in from them you will not do any better with your external customers.

Develop and reflect a vision for your business that will allow employees to develop this sense. Your vision should allow employees to see that their jobs are important. They must understand the reason why you are in business and how what they do everyday impacts the mission. There is a reason you are in business. Do you communicate that to your employees? Is that communication effective? Do you spend time on this during new employee orientation? Do you mandate on-going training to reinforce this idea? Make sure your employees understand why they do what they do, and how that impacts your customers. Incorporating that into their vision will allow employees to develop a sense of belonging to an organization that is trying to make a difference. Couple that with actions by senior management to bring that vision to life and you start to have an atmosphere that will be a major factor in attracting and retaining good employees. It is important to make sure that senior management is acting on these principles, as well. If you have certain skills or functions you say are important, but senior management is not practicing what they preach, then you set yourself up for cynical rejection by the employees. When you yourself do something, make sure the entire team, from top to bottom, is on board. If you do this you can create a dynamic workplace where people will know what they do really matters. Connecting the vision to each action can clarify work processes for people. By doing this you can make your workplace a place people want to be.

Do your employees brag so much about what they do that people come in looking for a job because they want to be a part of your organization? In our business (the firehouse), our employees (members) are our best advertisement for new members. Our members talk about what they do and how they make a difference. They introduce people to the fire service and get them to join. We are always happy to pursue more volunteers, but we have never in the history of our organization had to advertise or actively recruit new members. They come by word of mouth. They come because they see what we do. They come because they see how they can improve the world. Talk about a great way to save money!

In many cases it is difficult to get large groups of people with diverse backgrounds to develop a common sense of purpose. Fire departments are very successful in doing just that. Most fire departments operate in companies. (A fire service company is the same as a team in most corporations.) Internal competition is often used to foster a sense of belonging and purpose for the company members. Rewards may be offered (for example, the fastest drilling skills, most efficient use of hoses or ladders, etc.) but the key to promoting excellence is the desire of the team to be better than the other team. Good departments have made this into an art, albeit a very inexpensive art. They have monthly newsletters that recognize members. They have little plaques on the walls that recognize the firefighter of the month. At the end of the year they may offer an award for firefighter of the year and mention people with the most activity hours. This is pride—pride in doing the best you can. Pride in your team, for doing what you do better than anyone else. Success comes when the teams (companies) put their skills working as team members together to have a positive outcome at an emergency scene. How do you build for success in your organization?

Keeping an eye on the goal of your organization helps keep employees in the fire service satisfied. They will spend countless hours cleaning and maintaining equipment, training, and developing skills needed to be successful.

Have you ever heard employees say, "That's why they call it work?" Well, I have never heard that said about jobs at the firehouse. The social atmosphere around a firehouse is one reason that people get a great deal of satisfaction out of being a member. Good-natured ribbing, story telling, and folklore are all part of the firehouse culture. The social aspect, being able to give and take, makes the entire experience of being a firefighter a way of life, not just a job. The members themselves become the best advertisement for new employees, and are the reason no one wants to leave. Sure, some do leave, but very rarely do members walk away because they are unhappy with the experience.

In the firehouse, every member can feel that he or she matters, because we constantly reinforce the link between the most mundane tasks and our goals, which are to save lives and preserve property. Because members understand this connection, they know that they matter and are each important parts of the organization. Once the connection is made, retention is no longer a problem. Even though we don't pay them and ask the impossible of them, the retention rate is high.

How can you make working at your organization something every employee wants to make a way of life? You need to make time to meet the social needs of your employees. Having them work together on teams or activities produces social interaction. Making time for having fun and getting to know one another can go further than money in building employee loyalty.

Some questions to ask yourself:

- Do your employees understand their part in the big picture?
- Do they have a sense of belonging?
- Is their organizational pride evident in the behaviors they display?
- Is the level of interaction in your company adequate to help employees meet their social needs?

Rookie Firefighters Are Shown What Matters From Day One!

"If you don't have the time to do it right the first time, how in the world will you find the time to do it right the second time?"
                                                            —Chief Harkins

On a scale of one to ten, rate the accuracy of the following statements, as they pertain to your organization.

- We do a good job of training new employees.
- Our new employee-training program consists of much more than just teaming the new employee up with a long-term employee to teach them how to do the job.

- We spend whatever time is necessary to make certain that new employees understand our human resources policies and procedures and our benefits information.
- We strive to assure new employees are trained in how to do their job safely.

In many companies, new employee orientation consists of giving the employee a handbook, having them spend an hour or so electing benefits with the HR Department, and then teaming him or her up with an experienced employee to learn how to perform his or her job. Think about how you are training new employees and what message you are sending. How much of what is covered do you think the new employees will retain in your system? How much would you retain? Does that long-term employee know how to teach the new employee to perform the job correctly or is the new employee being taught "bad habits" that may have been developed over the years?

Let's face it, the longer it takes to train someone to do the job, the longer it is before they are being productive for the company. If you have read other chapters of this book, and haven't figured out ways to lower your turnover, motivate employees, and pay attention to details, then stop reading now because you will certainly disagree with the next statement. The long-term success of your business greatly depends on how good a job you do of training employees from the first day at work. The investment you make up front will be paid back many times over. That new receptionist isn't just answering the telephone and directing calls, that new machine operator isn't just grinding parts, and that new laborer isn't just moving concrete blocks. They are all YOUR COMPANY to your customers, suppliers, and the general public. How well they do their job is a direct reflection on you. The receptionist is the first voice they hear. The machine operator is a major player in the quality of the product you produce. The laborer can have a great impact on the efficiency and safety of your operation. How good a job you do of training/teaching/mentoring/developing employees will have a long-term impact on how they represent you, the quality of their work, and the safety and efficiency of their job performance. A famous hotel owner tells a great story about employee orientation. He tells the story of a waiter. On his first day he reports to work, goes to HR for a few hours, and then finds himself in the kitchen. When asked how he should learn the job they tell him, "Follow Betty for a few days and she will teach you the ropes." Well, in the first fifteen minutes our new employee is with Betty, she says to him, "This job stinks, the pay is awful, and the tips are not what you could get at the hotel across the street." Now the new employee is learning a lot about how the employees value the company, but not a whole lot about the company values. The owner pointed out to us that what Betty was doing was an actual orientation, because it was teaching the new employee how to behave. It was NOT in the way the company wanted, but in a way that would not serve the company well. The new employee could now easily find a place on the underground organizational chart, where the complainers and the critics have their own hierarchy and agenda. This is one reason that orientation is so important.

Thirty years ago when you joined our volunteer fire department, you were teamed up with a long-term member for your training. This was right about the time that self-contained breathing apparatus—commonly called an SCBA—was becoming a fixture in the fire service. If you wanted to "fit in" you wouldn't be caught wearing one. We were "smoke eaters." You were taught to find the refrigerator in a house fire, because the milk you would find there would soothe the burning and irritation in your throat and allow you to eat more smoke.

When you heard this from people that had been eating smoke for 20 years or more, you would assume that they knew what they were doing and you, too, would be a smoke eater. The techniques used were actually somewhat effective, with buildings built in the time before the widespread use of plastic. When you had wood and paper burning, the products of combustion were for the most part contained in the smoke. Fighting a fire without SCBA was possible by staying low, below the smoke, where the air was "good." In actuality, it was better, not good. With the advent of plastics, the characteristics of the toxic gases inside a fire changed. Many of the toxic gases given off when plastics burn stay low to the ground. No longer did you have the ability to see where the "bad stuff" was. To protect yourself the obvious answer was to bring the good air with you (in your SCBA). Older members were not advocates of this practice and fought the change for years (and some still do).

We realized years ago that we didn't have a great deal of turnover in our membership (and neither will you if you are following the guidance in this book). Investing time up front with new members (employees) has allowed us to develop in them the behaviors to perform their job as safely as possible. This is the time in their career when they are most open to learning the correct or accepted ways to perform tasks, represent the organization, and conduct themselves. In fact, new employees (members) expect to learn these things when they start a new job. Why not take advantage of that?

A gentleman by the name of William "Woody" Woodward developed our first formal Rookie Course. Prior to that it was very informal, and you could literally get sworn in as a member and respond to a fire the day you joined the department. Woody brought together a group of members to create a course to teach members the things they needed to know. It hit on many areas. First, they would need to understand our rules and regulations. These are the same as your policies and procedures.

Secondly, the new members would need to know how to perform assigned tasks safely. The first task they learn is how to get dressed. They are shown how they should look when they do their job. The personal protective equipment that firefighters wear is their lifeline, and the proper way to put it on, called "donning," may be the most important job they ever learn. We spend hours going over how the gloves should be put on so no spot of skin is exposed to the blistering heat inside of a structure fire. Every detail is drilled in repeatedly, so that the lessons are assimilated.

The rookies then move from the easier, safer tasks and work up to the more complex and hazardous ones. The training program was developed by breaking down each task into its individual steps. For each step the potential hazards were identified and the methods chosen to control those hazards were listed. This is a Job Safety Analysis (JSA). JSA is a very effective tool for both evaluating potential hazards of a job, and identifying and evaluating control methods for dealing with those hazards. These JSAs were discussed and developed with the input of the people currently doing the jobs (the long-term members). They were then used as guides by anyone assigned to train new members on how to perform certain tasks. The JSAs were also incorporated into our ongoing training programs and are the basis of our competency-based training. This gave us consistency in training, and actually helped our more experienced members become better mentors. Involving them in the process also helped change some of the long-held beliefs that they had about safely fighting a fire. It helped us, and continues to help us change with the times. This means that things learned in the rookie course are repeated in our weekly drills, and are tested under a stopwatch and watchful eye at the end of the year to make sure that the members have mastered these tasks. We have found that some of our best firefighters are the most lacking in attention to the details we go over in the rookie course. Years of experience have allowed them to develop shortcuts to get the job done quicker. To operate safely, we have to make sure that these shortcuts are not taken, so in our training we go over and over and over the basics about how to drive, how to dress and how to attack a fire. Sometimes our members complain that they are bored because we have repeated the same thing in our drills, but we remind them that our lives hang on these details. From time to time an experienced member (or a new member) may come up with a shortcut that is as safe and effective as the way we are operating. In those cases, we may well change the way we are operating to take advantage of the more efficient way of working.

Firefighting requires mastery of a number of skills. It is not reasonable to expect a new member to gain expertise in all areas overnight. Our rookie course provides an orientation, and we require completion of formal firefighting courses over the first three years of membership to become an interior structural firefighter. The rookie course starts small. The ability to identify equipment (and where it can be found) is the starting point. This is followed up by the proper usage of hand tools, appliances, ladders, hoses, nozzles, and power tools.

Adults learn by doing. It is said that you remember 10 percent of what you hear, 20 percent of what you see, and 90 percent of what you learn to do yourself. During our rookie course each new member gets his or her hands on every piece of equipment we own. Experience is gained using each tool during closely monitored training sessions. This allows us to assure that each new member can demonstrate the proper usage from day one. We strongly believe that the way rookies are originally taught is the way they will remember to do the job. Doesn't it make a lot of sense to make sure they are learning the correct way first?

It certainly may take some additional time (the first time) to set up your training and do it right. It is well worth the time you will invest, because if you do it right the first time, you won't have to do it again.

We know that we throw a lot at new members, and we know they can't possibly retain it all, so each new member is assigned a mentor. This is a person that is available to them to discuss any issues that have come up. Having a person from whom they can get answers and alleviate concerns without having to go to the boss has proven effective in addressing minor issues before they become major incidents. Mentoring has helped us retain members who otherwise may have walked away or become disgruntled. The mentor works with the new member at training sessions and emergency scenes to reinforce the lessons learned during the Rookie Course.

The effort we place on making new people members has been rewarded a thousand times over.

In your business, just like our fire department, the quality of your orientation and training process has a direct impact on your bottom line. How do you feel when your first call to a supplier gets cut off or you are connected to the wrong party? How do you feel when you open that package you've been waiting for only to find a defective part? What impression do customers get when they do business with you? Remember, first impressions are forever. You never get a second chance to make one.

New employee orientation will form a bond between your employee and your team, or it will be the first wedge in the war between "them and us." Rookies can join you or become part of that underground organizational chart of complaining and criticizing that plays such an important part in so many organizations. This is your chance to get another player on board that can help sell your company to the world.

Here are some questions to answer to help you improve your new employee orientation.

- Have we completed Job Safety Analyses (JSAs) on all jobs in our facility? Can we make use of these as a training aid for new employees?
- Can we formalize steps in this process to assure we are consistent in the information we present to new employees?
- Identify six things that could be done to provide a better new employee orientation:

Secret Number 10:

Training Is A Lifesaver!

"We can chart our future clearly and wisely only when we know the path which has led to the present."

—*Adlai E. Stevenson*

On a scale of one to ten, rate the accuracy of the following statements, as they pertain to your organization.

- Our organization has training programs for all levels of employees (from the line employees to senior management).
- Training programs in our organization are aligned with our vision/mission statement.
- Our training program is effective.

There has been talk over the last few years about the "learning organization." This means that the organization should be a place where mistakes are learned from and bad decisions and behaviors are not repeated.

To be a "learning organization" requires more than just talk. It takes a commitment to certain levels of training and education. It is important that a distinction is made between training and education. Training addresses those skills or practices that can be taught and practiced. Education addresses those behaviors that are learned more subtly. While organizations talk about being learning organizations, in reality, there is often very little linking the training they provide to their mission.

In many corporations the training departments are often the first to be cut when there are constraints put on budgets. Employee off-site training is often a primary target when the budget is tight. In addition, while lower level employees may be able to get training in such areas as computer skills and time management, less training is often available the higher up you go in the corporation. To really have a learning organization, training and education must be provided from top to bottom. Training is often not available for middle and upper management. As a matter of fact, in many corporate cultures across America, there is a stigma attached to training, so management may never get the benefits of new theories or the interactions with people from other firms. To be a "learning organization" everyone should be educating themselves all of the time, not just new employees learning about Microsoft Word.

Fire departments cannot afford to be lax about training. They cannot afford to repeat past mistakes, making it imperative that lessons are learned from each incident. Most fire departments address this by providing a rigorous and carefully planned training program. The past has been a good teacher to the fire service. Thucydides said that war is a cruel teacher. We can say the same about fire. Its' lessons can be deadly and unforgiving. It is by looking at their past that fire departments have become painfully aware that performance is linked to training and that training should be linked to benchmarks and expectations of performance. These in turn flow right out of the mission statement. These training modules will help realize the mission statement in the day-to-day workings of the business.

A mission statement is the map of where we as an organization are going. Many mission statements use words like "world class" or "industry leading" or "customer satisfaction." In fact, those high-sounding words are often not reflected in the day-to-day workings of the organization. This is because the mission statement is not really taken seriously. There is little pressure to actualize it via training and practice. If the mission statement were a "real" document, then training would probably be the highest priority of the organization. Read a company's mission statement and then look at its training department and training schedule. Is there a strong connection between the two? The fire department training officer is one of the most important people in any fire department. The training officer or training committee can make the difference between an excellent fire department and one that could use improvement. The mission statement of the fire department will show us what the goals and objectives are.

The mission statement at our firehouse encourages us to be a recognized leader in providing emergency services. Once we have said that we want to be a recognized leader, many actions will follow from that. From this mission statement we can see that our members need to be trained in certain ways. To actualize this, several levels of organizational commitment are imperative.

Our fire department has gone to a competency-based training system, or as we call it, "CBT." Instead of counting training and education as the time spent sitting in a chair listening to someone talk, we have identified a number of core competencies that our members must have. This training fulfills two purposes. First, it allows us to cover our mandated safety training. This encompasses government mandates and other requirements for annual safety training. The second purpose is to work on the skills that we need to be effective firefighters and EMS workers.

At the beginning of the year our training committee determines what objectives we must meet. For example, one objective may be getting the first firefighters off of a fire engine and into a burning building with a certain amount of equipment under a certain benchmarked time. Once this objective has been established this is written into the training plan. We then put it into our competency-based training. It would read something like this:

> "Successfully stretch a hose line to the front door from an
> attack engine within 30 seconds of arrival."

To complete this task for the CBT test at the end of the year, we must make sure that our training each week is aligned with achieving the results we expect to see. This allows us to regularly reinforce those skills that we have deemed to be essential. To recap, our weekly drill schedule is set up to allow us to work over and over again at coaching the skills and behaviors that we have deemed crucial. These skills and behaviors are addressed in our training objectives because we feel they are essential for us to achieve our mission. You can see how this example ties right into our mission statement.

Weekly training is done because we recognize that adult learners learn best when they have their hands on their lesson. It is not enough to merely talk about these things. We must practice them over and over. Our drills are often a repetitive series of exercises in which we reinforce those skills that might just keep us alive and help us do our job better.

- Is your organization a learning organization?
- What would it take to make your organization a learning organization?
- Are your vision and mission statements reflected in your training schedule and curriculum? What does this tell you?

We have shown you the ten secrets volunteer fire departments use to burn for success.

To help you achieve these results in your own organization we have included what we call "fire starters."

These fire starters will help light a fire under your organization and get you going in the right direction.

Firestarter 1:

Always Tell The Truth

"Don't find fault, find a remedy."

—*Henry Ford*

On a scale of one to ten, rate the accuracy of the following statements, as they pertain to your organization.

- Our company has a performance evaluation process in place that gets input from all parties who are impacted by the performance of the individual/team.
- The information we are getting on team/individual performance is timely, and is being used to help the team/individual perform at a higher level.
- The purpose of our performance evaluations is employee/team development, not to justify whatever raise or promotion should be forthcoming.

It is hard to tell the truth, and often managers and organizations shy away from it. It is a rare person in any organization that will call things what they are. Organizations, be they fire departments or businesses, are prone to being dishonest about issues that are painful to the organization. In some corporations, the "party line" is something given out by management and not believed by employees. This drives the wedge between "us and them" deeper. Dishonesty can be tolerated as long as it will hurt no one. In the fire service that is never an option.

Job/team performance should be evaluated on a regular basis. It should be evaluated to help each person or team continue to develop. It is a means to identify areas of strength and areas where improvement may be needed to perform better. How else will you be able to exploit the talents of your team and organization? That's right, exploit! Isn't that exactly what we want to do, make use of our strong points to outperform our competition? Performance evaluation should be an on-going process, not an event that precedes an annual merit review or raise. If you are doing a good job of evaluating performance on a regular basis, there won't be any surprises when it comes around to merit review time.

Often times we are not totally honest with our employees. There may be areas of improvement they could work on, but instead of addressing them, we hold off for fear we may hurt their feelings or step on someone's toes. This does a disservice to both the employee and the company because there is no opportunity to address the problems. We often talk about the value of team behaviors and being team players, but this does not alleviate us from the responsibility of telling the truth and dealing with problems as they arise. In fact, good team behaviors make it more important to attack problems early.

Here's a little exercise to try if you have employees who get merit reviews. Let them write their own review this year, and support what they write with specific examples. Compare what they write to what you would have written about them. Are they the same? If they aren't the same, you can be doing a better job of evaluating performance on an on-going basis.

Look at this from another point of view: If what the employee writes down is not even close to what you think of his or her performance, aren't there other issues that should be addressed? Employees and teams have a right to know where they stand. Even in an environment where a collective bargaining agreement (union shop) is in place, employees and/or their representatives need to know how they are performing and what needs to be done to improve performance. Managers in union shops need to know how their "team" is performing in relation to the expectations senior management has for them. Feedback (negative and positive) needs to be timely so mid-course corrections can be made. It is said that, "leaders don't have to make the right decisions, they only need to make them come out right." Thank goodness for that! It makes it a whole lot easier to be a leader.

From reading other chapters in this book, you certainly have gotten a sense of the importance of job performance in emergency services. You are also probably wondering how we evaluate job performance on a timely basis. Granted, it is not the time to evaluate someone's performance when there are flames "blowing out" the front windows of a row of stores, but there are many opportunities to evaluate that performance soon after the fire has been extinguished. In our department, this evaluation is called a critique. We critique every incident we have. This gives us a 360° view of the event, and allows us to identify what went well, what didn't, what problems were encountered, and what should be done differently the next time we have a similar event. It is extremely effective when done in a non-judgmental way. The purpose is well known by our officers and members, and the rules of conduct keep it from being a finger-pointing argument.

Chief officers in our fire department have found this to be a great learning tool because it allows them an opportunity to "see" what went on where they couldn't be. It is also a very useful training tool for junior officers and members. Not only does this reinforce why we perform certain tasks the way we do, but also helps them learn the strategic planning aspects of an emergency scene. Remember, there are sometimes flames rolling over our heads while we are trying to carry out our strategic plan!

How do we do a critique? Immediately following a fire, serious motor vehicle accident, hazardous materials incident or underwater rescue call, the Incident Commander will meet with the other officers present to talk about the event. This is a mini-critique, used to identify any major problems or actions needed immediately to safeguard the membership and organization. As soon as practical after the event we bring everyone together to talk about what went on from his or her perspective. The Incident Commander will usually start off the critique by talking about what he/she saw (saw, heard, etc.), what actions were taken as a result of that, what they thought went well, and areas of concern they saw. Other members and officers will add their perspective. For instance, the interior attack crew at a structure fire will add the conditions they encountered upon entering the building, problems they encountered in extinguishing the fire, oddities of the building construction, etc. The Incident Safety Officer(s) will give a report from their point of view. Once the picture of the incident is complete, the discussion turns to identifying what we should be doing differently. The critique helps us learn from our mistakes and helps us identify our strengths. That allows us to exploit those strengths at the next incident and put procedures in place to prevent the mistakes from being repeated.

Critiques are not the only performance feedback that is given. Senior officers are constantly providing performance feedback to junior officers. Junior officers are doing the same with the membership. Similar to any business, this feedback and development of action plans for individuals is used to help us in succession planning. Every business needs to have people ready to take on added roles and responsibilities. Championship baseball teams don't just pull someone out of the stands to bat for the pitcher in the bottom of the ninth, do they? To be successful in business or to win a world championship in baseball someone has to be able to step up to the plate and hit a home run. You can't expect that person to do that if he/she has never had the chance to do it before. Your life may not be on the line, but how well you/your teams perform has a major impact on the success of your organization. That in turn can have a major impact on your life and the life of everyone in the organization. The impact is certainly too significant to leave to chance. Make sure you have developed ways to evaluate performance on an on-going basis, so midcourse corrections can be made. Give your teams/employees every opportunity to succeed.

Look at the following to help you determine how well you are doing in evaluating performance.

- If someone asked him or her, every member of my team would be able to articulate how well we are doing as a team, and what specific actions need to be taken for us to be successful as a company.
- Every person that reports to me is able to tell me specific actions that he or she needs to take this week (this month, this quarter) to help the team succeed.
- We hold regular critiques to help us identify actions that need to be taken to be more successful.
- Critiques are always conducted after an unusual occurrence or event to re-assess our action plans and direction.
- Each team member is comfortable enough to provide suggestions about what should be done to improve the team performance.
- Each team member is comfortable enough to provide suggestions about what I should do to improve my performance.

Firestarter 2:

Keep The Team On Task

The future is here, where are you?

—*Gabriel Hevesi*

On a scale of one to ten, rate the accuracy of the following statements, as they pertain to your organization.

- Our workplace makes use of technology to be more efficient.
- Our company has developed/adapted to alternative ways of working to capitalize on the technology available to us.
- The use of virtual teams is fully embraced by top management, and is accepted as the way we do business.

The way we "do" work has changed drastically over the last two decades. Multinational imperatives, complex shifting markets, globalization, mergers and acquisitions, cost constraints, and an explosion in internet technologies have made it necessary to change the way we work. We have gone from traditional co-located teams to remote and even virtual teams. We have gone from individuals and teams all in one place to getting input and contributions from numerous people from different cultures that are geographically separated. How well is this change going in your company? Have you had problems changing the way you do business? Have you seen reluctance in your management to let employees operate in a more independent team-oriented environment? I assure you, you are not alone.

A virtual team is a work team whose members are not located in the same place. Members can be in many different countries and continents, can span time zones, cultures, and styles of working. Looking at this you might be saying, "Okay, I understand that most businesses today are making more and more use of virtual and remote teaming, but how does this apply in the fire service, and more importantly, what can the fire service teach me about this?"

The problems and issues surrounding working remotely or virtually are not as new as you might think. Fire departments have been working with virtual and remote teams for centuries. Like your company, they have struggled with some of the technology and communication issues, but they have developed a very efficient model you should look at.

In the fire service, virtual and remote teams are commonplace. An engine company performing an interior fire attack being directed by an incident commander at a command post is a simple example of a remote team. An emergency scene requires the coordinated activities of a number of remote teams and/or individuals to reach a successful conclusion. Our simple example above is actually not that simple. At the same time that the engine company is making their "attack," another team must be providing ventilation of the fire area to lessen the potential for backdraft explosions or flashovers that could be fatal to the engine company. Another team will be performing a search for potential victims. Emergency medical units will be setting up rehabilitation and treatment areas. Additional engine companies will be supporting the operation by providing a supply of water and back-up crews. A FAST (Firefighter Assistance and Survival Team) team will be assembled to provide rescue of fallen firefighters, should the need arise. Although these individuals and teams are not operating across large geographical areas or time zones, for the most part they do not have the ability to meet face to face and must use the same tools available to you to successfully operate remotely.

How do fire departments accomplish this? How can you achieve results in spite of the challenges of virtual teaming? Experts point to 5 issues that must be addressed:

1. designing and supporting the team
2. building the team's culture
3. removing barriers
4. trust, and
5. stress.

Fire departments are not successful by accident, nor will you be. Fire departments design the teams (we call them companies) they have so that they will succeed. The members are provided with or bring the necessary training and skills to the team. They know what they will be asked to accomplish, and are provided with the resources (fire trucks, tools, protective gear) needed to get the job done.

The team(s) you will use must be designed to succeed. What is the mission/goal of the team? What resources are needed to make the team successful? How should members be selected, trained, and recognized for their contributions? What support will be required? How will the team communicate with one another and with other teams?

Fire departments and other emergency service organizations have their own culture. It is defined by the members of the organization and may vary between the different companies in any one organization.

Team members bring with them their national culture, the culture of the organization they are a part of, and perhaps even the world as seen from their position. Team leaders and members need to be aware of cultural differences. A person's culture will influence his/her behavior. Understanding different cultural biases can help us better understand why people behave and act the way they do. Each individual's culture can be affected by their national origin, their function, and the organizations they have been a part of. For instance, Asian cultures typically place great importance on the group (collectivism), while in the United States we are much more concerned with the individual. Singling out an individual for praise in a culture where the group is most important can be detrimental. Job functions can give individuals different cultural attributes. Engineers are typically highly analytical people with a need for structure. Philosophers on the other hand are comfortable in a world where there are no absolutes and there may not be any "right" answers.

With the world shrinking it is important to understand how culture can impact the way a person sees it. There are no absolutes (i.e., a person of Asian descent may be very individualistic), but a basic knowledge of cultural differences may help us understand why some people act the way they do.

Based on the make up of the team and the goal, the team will develop its own culture in much the same way that corporations and other businesses do. Changes in culture take time and planning. This must be accomplished in short, measurable stages. Your firm's leadership should be aware of cultural differences and how these differences may impact team performance.

Fire department leaders must address barriers to successful team performance on a routine basis. Whether it's the "long time" members resisting changes in Standard Operating Procedures or budgetary constraints, the barriers must be knocked down for us to function as a world class organization. There has been resistance to many changes that were proposed in the fire service. Just as in all organizations, members who liked things the way they were erected barriers to stop changes that they thought would damage the fire service. The barriers were eventually broken down by demonstrating the effectiveness of the changes in helping to get firefighters to the "seat of the fire."

In most businesses there are a number of barriers that can impact how successful virtual teams can be. Is the organization supportive of virtual and remote teams? Who in senior management is championing the move to virtual and remote teams? This individual has a vested interest in the success of the teams. Does everyone have and know how to use the technology the team will be using? Are there cultural or technological barriers to effective communication?

We need to be aware that what we think we communicated may not have been received the way we meant it. In the fire service we use a technique called "echoing" to make sure our communication is clear. It is similar to paraphrasing in that the key points of the message are repeated back to the sender to assure the communication was received as it was intended. How can you make sure your communication is clear?

Trust is the most important issue in emergency services. Team members must have trust in one another or they won't put their lives in each other's hands. Without trust a successful outcome to an emergency incident is doubtful. Trust in an individual or team is based on performance, consistency, integrity, and a demonstrated concern for others. Teams and individuals that are competent to perform the tasks at hand, are consistent, follow through on their commitments, and treat others as they would like to be treated, inspire and maintain our trust.

Building trust in a virtual environment is more difficult than when we are co-located. This is because many of the cues we use to determine if we trust someone are lost. Typically, we start to make judgments on trust once we get to know people. Their words, deeds, and actions all play a role in building trust. In a virtual environment, we need to find ways to build trust between members. If you have virtual teams or are putting them together at this time, consider the benefits of having the team members meet face to face for a kickoff. Have them spend their time together doing activities that will build trust. In the fire service, we use our training sessions and social activities to help individuals build and maintain trust in one another.

Trust is like money. It is harder to get than it is to lose. Things like punctuality can build or destroy trust. Not keeping a promise or missing a deadline is deadly for maintaining this kind of relationship. Over-promising and under delivering can kill trust at any stage of a relationship. When you are remote or virtual it is important to remember that every email you send is your handshake and every telephone conversation or Instant Message you send is your word.

Stress is part of the world we live in. It is a condition in our workplace today. The same is true in the volunteer fire service. In that world, like in yours, we are asking individuals to respond 24/7 without much time to reflect or think through a situation. Very few (if any) individuals operate in an environment that is stress free. There is stress from our working relationships, changes in technology, and all of the constant changes going on in our lives. The key to addressing stress then comes down to controlling it. Clear ground rules addressing conduct, moral, and ethical issues can help employees cope. Wellness programs (smoking cessation, blood pressure screenings) can help. Physical conditioning through regular exercise and a healthy lifestyle can help you and your employees manage the stress that will be present.

Virtual and remote teams are becoming a normal part of business. Performing successfully in spite of the challenges of working apart requires leaders to properly design their team and its support infrastructure, build and sustain the team's culture, remove barriers, maintain trust, and address stress. How well are you doing this?

Do your team members:

- keep their promises even if circumstances change?
- inform others well in advance if they will be late with deliverables?
- share new ideas with one another?
- respect the expertise of others?
- focus on results?
- do what is in the best interest of the team?
- do the right thing, even in a crisis?
- stand behind the team and its people?
- know how they impact other team members and other teams?

Firestarter 3:

Use The Right Tool For The Job

"The employer generally gets the employee he deserves."
—Sir Walter Gilbey

On a scale of one to ten, rate the accuracy of the following statements, as they pertain to your organization.

- Our company has processes in place to assure we are continuing to move in a positive direction.
- Our company has systems in place to assure that our processes are working.
- Our company has the right people in the right places.

It is said that leadership is doing the right things, and management is doing things right. This chapter will discuss doing things right (managing people and processes). Speaking strictly from a management perspective, the job of a manager is to manage people and processes. These two responsibilities go together. Managing people means you get the best out of each employee by doing two things. First of all, you must have the right employees in the right jobs. Second, you must keep them motivated and on task so that they do the best job possible.

Managing processes goes hand and hand with managing people. Good people will not be motivated if they work in a company that has no processes or bad processes. A bad process can crush even the most enthusiastic employee. If you have ever visited a government office where there are long lines, confusing processes, and complaining customers you are likely to find one more thing. The employees behind the counter are often rude, indifferent, and easily angered. It is not hard for us to imagine that they did not start out this way in the job. I am sure on their first day of work they were as enthusiastic and motivated as anyone could be. But working as part of a process that does not work will quickly frustrate and anger everyone, both the customers and the employees. It probably does not take long for a new employee to learn that in a place where few things work and you get no good feedback, the best thing is to do your job as minimally as possible and make sure your lunch hour and breaks are stretched out as long as they can go. A bad process will eat the best people. On the other hand, a great process will not work as well as it should if the wrong people run it.

The trick is to build both morale and the process together so that enthusiastic employees can operate inside of rational processes to make things work. Both of these tasks are important. In many companies there are tasks that are dependent on one particular person. He or she is the only one who knows certain procedures, can find certain items or can authorize certain exceptions. These abilities make this person indispensable, but at the same time it creates a potential bottleneck in any process because if this person is sick or busy, the whole process must come to a halt. Often times the people in these positions are your finest, hardest working, and most loyal employees. That's how they got to be so invaluable.

Office automation has helped speed up all work processes. No matter what field, the increasing use of automation demands that processes be developed that can then be automated and kept track of. Work must be rational and streamlined. We can no longer rely on that single individual who knows all and has been on the job forever. Systems must be put in place that can operate all of the time. This transformation of work is part of a manager's job. By developing such processes, we help our employees do their job. Hopefully, streamlining the process will remove obstacles from the paths of employees. These new process should improve employee performance. Hopefully, these changes will come with a set of clear expectations so we minimize frustration.

The other half of the manager's job is to motivate and train the right people. Having the right people in the job is half the battle. It is important for a good manager to deal with problem employees in a timely manner. Having a problem employee can impact the other members on the team if actions are not taken quickly. There are many ways to remediate these employees. Where possible, getting rid of the employee is a solution. In other cases there may be mitigating circumstances, such as unions and seniority issues, that make it impossible to terminate or transfer these employees.

Every manager that has ever come into a new situation inherits the existing employees. They also inherit all of the problems and difficulties that go with these employees. Luckily, they inherit all the talent and expertise that these employees possess, too. When a new manager steps into the job, he or she must take stock of where he or she is. An initial determination of strengths, problems, and areas needing improvement must be made. Where problems are identified the manager must identify if the problem is in the process or with the employees. Is the problem systemic or personality driven? Which factor(s) must be remedied for the job to get done better?

In today's environment, many companies are going to an integrated computer system that unifies all of their information in one place. This is sometimes called an ERP (Enterprise Resource Planning) and is used to link together all of the business processes in the company. When this happens, managers may find out that before they can link all of the business processes that the whole workflow needs to be reengineered. To link all the processes in the system, the whole process has to be revamped. In these days of information transformation, processes are under more and more scrutiny as they are linked together and make use of similar information resources. This also means that the people involved have to learn to do things a different way. They need to rethink everything and retool a great deal. The processes depend on the people and vice versa.

A volunteer fire department faces a particularly difficult challenge in this area. We do not get to "hire" from a pool of the best applicants. Because it is such a huge commitment of time and because it is based on a peculiar desire to serve, we cannot always select who comes to us. We have to do the job with whomever we can get. While it would help the smooth operation of our organizations to have more members with better management skills and experience in running things, it is often not the case. That is often the case in your business, as well. When unemployment is low, job applicants may be scarce, and you may have to make do with what you have.

Once you realize your choices of applicants are limited there is a rational next step that most organizations take. They build as foolproof a process as they can so that almost anyone can understand what is required, what must be done and what is totally unacceptable. During the later years of the Vietnam War there was a draft and large numbers of applicants (draftees) were taken into boot camps. Every process, from getting dressed to brushing your teeth to making your bed, was so highly detailed and managed that it brought people quickly into a system where they learned what was expected of them. Which buttons on your uniform had to be buttoned and in which order, and which part of the rifle had to be assembled first were not a matter of personal preference, but painstakingly laid out and explained. No variation was permitted.

Similarly, fire departments and ambulance corps outline and benchmark every single process they can. For example, every firefighter wears gloves. These gloves are inspected on a regular basis. There are certain defects or compromises in the material that will render them out of service, and the firefighter will get a new pair. This is specified exactly and not left up to the opinion of the member. Almost all processes in the firehouse are like this.

Matching the people with the process is the second great task that a manager has. Everyone has his or her strengths and weaknesses. The trick is to put a person in a job where their strengths are magnified and their weaknesses are minimized. Some people are great at building systems but don't motivate the people below them to buy into those systems. Other people may be great motivators but not very systematic. The first person might do well in operations or finance while the second may do well in HR or sales. The trick is to have each person in the place where they can do the best possible job for the organization. Successful organizations know this lesson well.

Sometimes the best firefighters are not the best officers. Someone could be great at attacking a fire and show leadership and courage inside a burning building but that does not mean he or she has the skill set to manage a large and complex organization. The same is true in other businesses. The best employees may not make the best managers. In fact, taking a good employee off the line and making them the manager could hurt the organization in two ways. The individual may not have the skill set to be a strong manager, and you have lost a strong employee from the line.

Once the processes are in place to insure that the job will get done, we must select and motivate our employees. The motivation and buy-in of the employees can make or break the system we are going to set up and run. How do we insure that motivation and buy-in?

If an employee is in a job that they are not suited for, he or she may not fail but he/she will no doubt be unhappy and frustrated. A Ph.D. in German philosophy may be a star in the classroom but be a miserable truck driver. We might guess that the same thing may be true about a great truck driver. So the trick is putting people where they can make the best use of their talents, can be most successful, and have the least chance of failure.

One way to insure an employee's success is to make sure that he or she is evaluated on a regular basis. This evaluation needs to be honest and direct. Good managers know their employee's strengths, weaknesses, preferences, and so on. They ought to know if any life events, such as illness in the family or marital problems, are impacting his or her performance. They ought to know the aspirations and time lines that the person is operating under. They should know the goals the employee has set for his/her life both at work and outside of work.

A performance evaluation is an opportunity for a conversation about how the employee sees the company and what his or her role is in it. It is a chance for the manager and employee to exchange views on where the process could be improved and what needs to be done to make this happen.

The evaluation should include both praise and suggestions for improvement. It should include an honest evaluation of the strengths the employee brings to the organization, and ways that those strengths can be capitalized on (for the benefit of the employee and the organization). There is room for improvement in everyone of us, just as there is room for continuous quality improvement in all of our processes.

Honesty with an employee about his or her weaknesses is a great gift if it is done correctly. The reason why Freud thought people should be in therapy is that he found out that most people find it very hard to be totally honest. They need an objective person to nudge them along and probe into areas they might be reluctant to explore. It is the same with our working lives. Many people are not clear about their weaknesses and have not told themselves the truth about where they are and what negatives they bring to the job.

An honest performance evaluation gives both parties an opportunity to explore what can improve and how the employee can best work in that environment. There will no doubt be resistance to such conversations, but it is important for the organization that this step is taken. The employee can offer valuable feedback here, as well. Because they are inside the process and work with it everyday from a different perspective, they can tell the manager things he or she may not know about the process. This can show the manager where the process is in need of improvement or overhaul.

So the manager wants to talk honestly to the employee about where there are areas of concern and improvement and the employee should be talking to the manager about areas of concern and improvement for processes. Both must trust each other to be honest enough to say what the problems are. If one or both of these parties is not honest, the communication process will break down. Either the employee will not be aware of areas of concern or there will be parts of the process that are not working correctly that the manager needs to be aware of.

Our firehouse recruited a retired gentleman a few years ago who drove a bread truck most of his life. When we voted on his membership, some of our members thought that he was too old to be of use inside of fires or on ambulance calls. He has turned out to be an asset to the department, because he comes to the firehouse every day. He empties the garbage cans. He sweeps up. He now heads our sick committee and keeps up on those members who are home ill. He buys the donuts every Sunday and handles a good number of other jobs. Someone who is riding the ambulance all of the time may look at this fellow and say he does very few fire and ambulance calls and therefore is not helping us out. In truth, what he does is take the burden of those odd jobs off the rest of us, and he does it well. A good manager recognizes these contributions and rewards everyone for them.

How are you managing your people and processes? Let us try a few exercises.

- Have you sat down with your employees who execute processes and ask them how it is going and how the process can be improved?
- Is the right person in the right job in your office? If you are not sure of the answer to this question, ask him or her.
- How could people and processes be better matched for a more efficient office/operation?

The Best Way To Predict The Future Is To Invent It

"We must adjust to an ever changing road…while holding onto our unchanging principles."

—Unknown

On a scale of one to ten, rate the accuracy of the following statements, as they pertain to your organization.

- Employees at all levels of our company make changes to improve the operations under their control.
- Managers/Leaders in our organization embrace change as the road to improved results.
- People at the appropriate level make decisions rapidly in our organization.

It is a myth of the information age that change is something new to business. Change has always been at the heart of the process. Even the manufacturing plants of fifty years ago were undergoing constant change, constant improvement, and constant modification of processes. Although the speed of change may not have been as rapid, the stress of change was still apparent.

What technology has done is to speed up the pace of change. While there has always been change, it may not have been as apparent to everyone. This increase in the speed of change has impacted every aspect of the organization, but the most important impact has been on the management structure.

The old style of management was the hierarchical pyramid with the big boss on top and trusted middle managers giving orders to the trusted lieutenants near the bottom of the pyramid who got the job done. There was a lot that was good about the hierarchical structure, but it had a major shortcoming. For any decision to become a reality, the information had to go up the pyramid. A decision would be made and then an order would come back down the pyramid. This style of decision-making is fine as long as the flow of information moves at a moderate pace and the decisions that it requires can be implemented over time.

What the information revolution did was increase both the volume of information as well as the speed at which information travels. The volume of information increased because the ability of the technology to carry and save it increased. In the days of ancient Sumer, when each piece of information had to be written in a specially baked mud brick, people would think twice about what they would write. In the Middle Ages, when each book took a single monk an entire year to write and illustrate, not all stories were deemed worthy of being preserved. As the technology to make and store material got cheaper and became more available, more information became part of the equation.

Speed has always been important for business. The Titanic was wrecked attempting to set a record for a transatlantic crossing. Speed has always sold because people had goods that were perishable. Speed has always sold because you could beat your competitors to market. The "fastest" has always been one of the benchmarks against which others were judged so when speed became widely available, it is easy to see that businesses are adapting to the new pace.

These changes brought about new theories of management. In the old pyramid structure the underlings would never act without first "kicking it upstairs" to see what the boss would think. The biggest sin you could make in the old hierarchical structure was to exceed your authority. That was seen as a threat to the power structure and often had severe consequences. As the flow of information increased it became more and more difficult (in many cases, impossible) to check upstairs every time a decision had to be made. In time, it became apparent that empowerment was something that was needed for businesses to be competitive. For businesses to keep current, they had to stop worrying about the chain of command and spend more time making the process flow cleanly.

This has allowed organizations to more rapidly adapt to change. They can now see where things need to be done and react more quickly to capitalize on the opportunities that are being presented. In the old pyramid structure, if someone down at the bottom had a good idea about how to make the business more efficient, it might have taken a long time for the idea to go up the chain of command. As it traveled, each person who passed it along got a little credit for it. In today's new, more interactive structure, it is possible to get a good idea to top management in a very short period of time. Sometimes it only takes the time you need to write and send an email!

Fire departments have to be very flexible organizations. They must listen to the feedback of the people doing the job and be ready to change their whole strategy based on a single piece of information. Let me give an example.

One of the first decisions an Incident Commander (IC) makes is whether a fire will be fought as an offensive or defensive operation. If the fire is going to be fought as an offensive operation, people are committed to going inside the structure and aggressively battling the fire. If the building is abandoned and the fire is advanced, the IC may decide to do a defensive operation where he or she assumes that the risk to firefighters is greater than the reward of putting out a fire in a property with no real value. This kind of risk assessment is crucial to any type of firefighting.

A fire may begin as an offensive operation. The firefighters are committed to going inside and battling the blaze. Once inside, however, they may discover some information that would be of use to the IC. Let us imagine there is a basement fire in a house. The firefighters are inside a very hot and active fire when they discover several large fuel tanks stored there, already in the path of the fire. If the fire cannot be extinguished right there, a decision may be made to remove the firefighters so that their lives would not be risked for a building in which there is no life hazard to any civilians. This piece of information given to the IC will change how the fire is fought almost immediately.

The ability to rapidly change direction is one of the most important skills in the information economy. To "turn on a dime" gives us the ability to attack new markets and set new directions. One of the most striking examples of this is the change that took place at Microsoft. Bill Gates made his fortune on operating systems and software that sat on individual computers. He bet that people would buy his software and run it on their own computers. When the Internet came along, so did the idea of distributed computing. This meant that the software would not be on any one computer but would be accessed by the 'net. Ten years from now, there will probably be little on our computers except the ability to hook into the network where everything resides. Bill Gates took his company from a focus on software that resides on machines to an Internet company with MSN network and all of the things that go along with that. The ability to change the direction of a company as large as Microsoft shows it can be done.

How attuned are you to managing change in your organization? Rate yourself in the following areas:

- Do employees at all levels of your organization have the ability to make changes in the operations under their control?
- Are employees rewarded for making decisions?
- Are we rewarding the process or the results? (Maybe we should be rewarding excellent failures and punishing mediocre successes—a bold statement that came from a participant, Phil Daniels, in a Tom Peters seminar on "The Work Matters.")
- Is there accountability to help employees, supervisors, managers, etc. make decisions, or is the safe way (pass it up the line) the way you are doing business?

Get The Best Out Of Your People

"Far and away the best prize that life has to offer is the chance to work hard at work worth doing."

—*Theodore Roosevelt*

On a scale of one to ten, rate the accuracy of the following statements, as they pertain to your organization.

- Our employees know their real strengths and weaknesses.
- Our organization knows how to bring the best out of every employee.
- Our organization maximizes the strengths of our employees.

In many corporations there are armies of disenfranchised workers who no longer feel part of the team. For a variety of reasons they are not positive attributes for the organization but pockets of negativity and discontent that work to thwart every new effort and initiative of the corporation.

They have developed a "them" and "us" viewpoint that guides all of their thoughts and actions. They have a place on the underground organizational chart where there is a hierarchy of complainers and critics. There should be a chart for such a sociological organization because there is often a CEO of complainers, a boss of critics, and a chief of complaining. Their boardroom is the coffee pot or water cooler. Their "employees" are the other cynics and critics. There is a firehouse near ours that also has a Sunday morning work detail. There is a coffee pot there and that firehouse has a very highly organized underground organizational chart. The CEO of complaining spends Sunday mornings complaining, critiquing, and generally undermining any efforts made by the organization. Now, this CEO of complaining does not try to change the organization and monthly meetings or serve on any committees that would make things better. He just brings everyone to the point of anger by giving, often in humorous ways, a commentary on the firehouse's activities. Anything that the chief does is brought under the harshest scrutiny. No positive solutions are offered and no effort is made to communicate a better way of doing things. The underground organization chart has its own ends, and they are often very different than the ends of the organization.

Divisions like these in the workplace were a force in the formation of unions. The battles for unionization came about because workers no longer believed in the good will of management. Many workers who hold a "them" vs. "us" attitude are not toiling for the greater good of the corporation. They are putting in their time until five o'clock on Friday, when they can leave and do something they consider worthwhile and important.

There are countless reasons why this happens, but one of the most important reasons is because these workers feel undervalued by their supervisors and/or top management for the job they do. If you walk through many large offices, you can find people playing games on their computer, cruising the Internet, making personal phone calls, and just passing time. They no longer feel (or never felt) connected to the goals and mission of the corporation. They feel isolated and undervalued. How does this happen and how can it be prevented?

One reason why this happens is that employees often learn that their weaknesses are more apparent to management than their strengths. Let us take an example. Bill is a new employee of X Corporation. Bill figures out a way to improve sales in the Southeastern region. So Bill, with his own time and with the enthusiasm of a new worker, writes a document demonstrating the problem in sales in that region, the data why he believes it to be correct and a solution that will make Corporation X a good deal of money. He submits it to his boss in a plastic folder and anxiously awaits the feedback for his initiative. After some days, Bill's boss comes to his cubical with the report. He tells Bill the following:

"I got your report and was anxious to read it when I discovered that your writing style really needs work. This first page starts out in a pretty unfocused way and really took a long time to get to the point. I am a very busy guy, Bill, and I don't have the time to stumble around with stuff like this. You need to work on your writing and organizational skills. I did not realize they were so weak when I hired you. Take this report and think about it some more."

Now let us imagine that Bill's boss is right. Let us imagine that Bill is both disorganized and could use help on his writing. Is that the real lesson that the boss just passed on? No. The real lesson that the boss taught Bill is not to take any initiative. The real lesson that Bill learned is not to go above and beyond for the corporation. The effect of that feedback from Bill's manager was to kill Bill's drive to improve sales and find other ways of bettering Corporation X.

In every corporation, there are dozens of Bills who know how to make things better, who know how to improve processes, and who know how to save and make more money. But they have learned to be quiet. They have learned that the corporation does not want to hear what they have to say. Although the corporation talks about being a "learning organization" and talks about how "every employee is valued," in reality they value not rocking the boat. They value dotting their i's and crossing their t's rather than improving business, manufacturing, and sales processes. The scenario above may not happen at your company, but don't bet the ranch on it unless you have looked. Every day businesses are killing so many good ideas that one has to wonder whether managers do this on purpose to stifle employee motivation or if they are just woefully unaware of the basics of human psychology. Perhaps you should spend some time making sure your company managers are not ringing the ambition out of their staffs. Wouldn't it have been more helpful if Bill's boss got past the shortcomings of Bill's proposal and saw the potential upside to his ideas? It was Skinner's behaviorism that taught us that if a behavior is rewarded with positive reinforcement or feedback it would continue. If a behavior is met with negative reinforcement, that behavior will be stopped or, as Skinner said, "extinguished." If we walk down the aisles of the major corporations we will see armies of human capital playing solitaire on their computers because they have learned that doing nothing is really what the corporation wants. The drive many of them had has been wrung out of them by their managers. This is an old story that has been written over and over. Is it being written again today in your company? Volunteer fire departments cannot afford dead wood. They can't afford to have members who are just putting in their time. Every ambulance call takes two or three people and throws them into a situation where someone could die if everything is not done exactly right. A fire engine pulling up to a structure fire may have a driver and three or four other firefighters. Every one of them has a job to do that is crucial, absolutely crucial, to the success of the initial attack on that fire. One weak link can break the chain. If we ask someone to hook up the hose to another truck to insure a constant supply of water to the attack team on the nozzle and that person fails to hook up that hose properly, the whole operation can grind to a screeching halt. The immediacy and finality of the situation does not allow them the option of having people who are not with the program. That is not to say that there aren't periods of time in the life of a fire department where members should air their views on how things could be done better. But problems must be dealt with quickly and effectively. The consequences of not dealing with them can be deadly.

Most fire departments are close-knit and very clannish organizations. They are often times not friendly to new members or strangers. That is because the organization cannot tolerate people who are there to coast. They are the "goldbrickers," as we used to say. Fire departments, as a result, are good at managing weakness. They have figured out how to encourage people to take the initiative and work for the organization rather than against it. They have learned not to kill the goose that lays the golden egg. How do they do this?

Fire departments, like most organizations, have many layers. As a result, members have the ability to do one job, a job they do well, and they are not required to do tasks they are either not good at or don't like. Fire departments, like most organizations, are not filled with superstars. There certainly are some, but what fire departments have done well is find ways for people to help the organization without having to be a superstar. What they have excelled at is making their members feel that they are superstars.

For example, there is a person in our fire department who is an excellent mechanic and driver. He loves to work on brake pads, adjust transmissions, and fix what is ever wrong mechanically. He does not however like to ride in the back of the ambulance when someone is bleeding or vomiting. That is not his "thing," but his value to the organization should not be denigrated because he doesn't want to perform that single task. There is a need for mechanics in a volunteer fire department just as there is a need for Emergency Medical Technicians (EMTs). You could not run a department with all of one and none of the other.

In a good department the mechanic will be rewarded for his skills and not punished for those things that he does not do well. In our fire department there is a member who only cuts the grass. This is an important job that needs to get done. We should honor that the same way we do the person who drives the fire truck or fights fires.

Every one of us has our strengths and weaknesses. A great organization is one where our strengths are reinforced and we are not punished for our weaknesses. This is not to say we should ignore the weaknesses, but we must realize where we can make the greatest contribution. Said another way, if we can identify the strengths of our people, leverage these strengths while still being honest about areas needing improvement (not making them the focal point), the benefits to the organization will be maximized. If you could energize those employees instead of alienating them, you would have a killer team. There are great coaches in sports who can take mediocre talent and come up year after year with winning teams. This is because they capitalize on their players' strengths and don't play to their weaknesses.

- Do your managers hold regular performance evaluations with individuals to identify ways the individuals' strengths could be maximized, or are your evaluations focusing on areas needing improvement?
- For each employee that reports to you identify three strengths:
- For each identified strength identify a course of action that would lead your unit/division/corporation to superior results:
- Identify six ways that you can keep your focus on the positive contributions of your staff:

Firestarter 6:

Take The Risk
Where There Is Benefit

"A ship in the harbor is safe, but that's not what ships were made for."

—*Grace Murray Hopper*

On a scale of one to ten, rate the accuracy of the following statements, as they pertain to your organization.

- Our company has identified the risks that we face.
- Our company does a good job of prioritizing what we should be doing from a risk management standpoint?
- Our company is in compliance with the regulations of OSHA, DOT, EPA, SEC, etc.?

You can't be successful without taking risks, but that doesn't mean you have to take unnecessary risks. Every business, whether they employ a Risk

Manager or not, must manage their risks or face the fact that these risks could put them out of business.

To understand how to manage risk, we must first define it. For our purposes, risk means the chance of loss or injury and also the probability of that loss occurring. Three interrelated factors can be used to determine risk:

-the **probability** that an undesired event might occur

-a **harmful** or **undesirable** consequence

-the **severity** of the harm that might result

The term "risk management" refers to any activity that involves the evaluation or comparison of risks and the development of approaches that change the probability or consequences of a harmful action.

Risk management is a process used to identify and evaluates risks. The process also identifies, selects, and implements control measures that may reduce the risk.

Let's take a closer look at the steps in the Risk Management Process. To manage a risk, it must first be identified. Once identified, the impact on the organization or individual must be evaluated. A method of managing the risk must then be determined. For example, driving a vehicle poses a risk of motor vehicle accidents. Perhaps the least significant, but an obvious consequence of a motor vehicle accident is property damage to the vehicle. Having identified this risk, let's evaluate the impact. To reduce the financial impact, most companies and individuals purchase an insurance contract to help pay for the repairs.

In addition to the property damage to the vehicle, we need to consider that it will take the body shop a certain number of days/weeks to repair the vehicle. This leaves the company/individual without the use of the vehicle. This can impact our ability to deliver goods to our customers or drive the kids to soccer practice. In emergency service organizations it can impact the service we are able to supply to the community. What can be done to reduce this risk? Many companies and emergency service organizations have more than one vehicle. If there is only one vehicle of that type, the loss of its use is more severe than if there were two or more. In emergency service organizations there may be mutual aid plans to assist agencies in handling emergency calls when they are unable to respond. In our department, should we have both of our ambulances out of service or on other calls, we will have personnel respond to the scene and render care until a neighboring department can respond to handle the transportation and care. Your organization may want to consider a rental vehicle or make other arrangements to minimize the impact to your customers.

This example took us through the risk management process. As a review, the steps in the process are:

> Identify the risk(s)
> Evaluate the impact
> Select methods for managing the risk

One other step should be added to this. Once we have selected a method to reduce a risk and have implemented it, we need to evaluate how well the control method is working, and whether or not it is still appropriate. The control method is the tool we use to reduce risk. This is what makes risk management an on-going process. How we treated a risk twenty years ago (or even last year or month) is not necessarily how we should treat the risk today.

The simple example above shows the need for risk management in every business including emergency services. Fire and EMS services were set up to help communities manage risk. Fires a hundred years ago were often conflagrations burning down major portions of towns or cities. Communities needed a way to manage the risks posed by fires, etc. Fire departments were one method used to manage the risk of fire. Building codes and municipal water systems are other methods that have been used. Similarly, the EMS service was used by communities as a method to manage the risk of injuries, and sudden unexpected illnesses to its residents.

There are four methods of managing risks:

Avoid the risk   Retain the risk

Transfer the risk   Prevent the risk

Avoiding a risk can always be done, but it is often not a practical approach to take. Exposure to tainted blood is a risk in providing emergency medical services. To avoid this risk totally, we can simply not provide emergency care. This is not very practical if we are going to provide this service to our community.

Some risks are so great that they require radical solutions. For example, a number of insurance companies recently announced that they would no longer write medical malpractice insurance in New Jersey. It was their opinion that the risk to their bottom line (stockholders) was too great to continue to provide coverage for health care professionals in this state. By no longer writing this insurance coverage they are avoiding the risk.

Many large corporations make a conscious decision to retain the monetary risks that many individuals prefer to transfer to others. A large corporation may self-insure their automobiles for physical damage (collision). As individuals, we may also retain risks. If we have an older car, or one that is paid off, we may retain the risk of physical damage to that car by not purchasing collision insurance. When discussing risk management methods, retaining a risk means that a risk has been identified, and we are willing to assume and retain that risk. In fact, any risk we haven't identified and controlled by some other method we retain.

Insurance is the most common method of risk transfer. By buying a homeowner's insurance policy, you transfer the monetary risk of loss to the insurance company. With prevention, we may use a combination of items to either prevent a person from becoming injured or limit the possibility of an injury or accident.

In actual practice, a combination of methods is often used to manage any particular risk. Let's take a look at our vehicle that may become involved in an accident. We identified that there was a risk of damage to the vehicle. As an organization, we decide that we need to *transfer* the monetary loss that would be associated with this to an insurance company. We may also decide that to get this insurance at a more reasonable price, we will *retain* the first $1,000 of any loss. To help *prevent* a loss from occurring, we will train all our drivers in the techniques needed to safely operate a motor vehicle, and we will set up a maintenance program to make sure the vehicle is kept in proper working order. If there was a particularly hazardous intersection leaving our premises, we might *avoid* that intersection by requiring all drivers to use the alternate exit.

The specifics of risk transfer and retention of risk are beyond the scope of this book, but we need to spend some time talking about prevention. Emergency service organizations are risk management organizations. Their origins were in suppression (minimizing a loss once it occurred), but they have evolved into prevention organizations. Fire prevention education, building code enforcement, and accident reduction programs are but a few of the activities that emergency service organizations are involved in. These activities focus on prevention of loss caused by the risks you face.

Loss prevention is also a major focus for emergency service organizations just like other businesses. Whether operating at an emergency scene or at the station, safety of the members and the public is paramount. (Preventing accidents, injuries, and illnesses also helps us get the best possible pricing on risks we transfer to our insurance carrier). Emergency service organizations must deal with OSHA, DOT, EPA and other government agencies, just as you do. Our department has a safety officer and there is a safety committee made up of management personnel and labor.

The committee is actively involved in development of our "Normal Operating Guidelines." These are the guidelines that we follow to make our operations as safe as possible. These guidelines may determine how we do much of our business. They are comparable to a Job Safety Analysis or other operational tool you may have used to identify methods to control injuries or prevent accidents in your jobs.

At each of our emergency scenes we have a Safety Officer. This individual has responsibility for monitoring all activities. They have the power to stop the operation if members are operating in unsafe manner or if unsafe conditions are identified. Of course, this action is communicated immediately to top management at the scene (the Incident Commander). Our Safety Officers have training specific to being safety officers and have been very successful in helping us prevent injuries under rapidly changing conditions.

Prevention should be a key in your risk management plan. Preventing accidents makes sense—dollars and cents. Companies are often associated with "bad" things that happen to them (Ford and Firestone for tire-related SUV rollovers, Exxon for the Valdez oil spill, Enron for being Enron). Identifying the risks you face, avoiding the ones you can, preventing as many as possible, and looking at the risk transfer and retention options available to you will help ensure that your long-term success is not ruined by accident.

Here are some questions to help you evaluate your risk management options:

- Identify the risks that you face as a company (hints: look at business interruption, product/property losses, vehicles, employee safety, customers/public on-site, etc.):
- For each risk identified, identify a control method or methods that you will use to offset the risk:

- Where you have already implemented control methods in your company, how well are they working?
- Consider having an outside person help you determine if you face risks that you have not yet identified (if you haven't identified and implemented controls, you have retained the risk).

Firestarter 7:

Promote People For The Right Reasons

"There is something that is much more scarce, something rarer than ability. It is the ability to recognize ability.

—*Robert Half*

On a scale of one to ten, rate the accuracy of the following statements, as they pertain to your organization.

- Our company does a good job of recognizing ability, and preparing people with the ability to take on added responsibilities.
- Someone in our company is prepared and able to step up and fill in for each and every key player in our organization.
- Our managers/leaders promote, cross-pollinate, and have subordinates take lateral transfers to explore different areas of the organization.

It is not uncommon for someone not to want to share everything they know about their job. By being the only one that knows things, a person has greater job security. But what if that person did not come in tomorrow? Where would the company be? We must make sure that we have prepared someone to step in and do the job if this were to happen.

In today's environment, where organizations are competing in a global arena, the need for succession planning is heightened. To be successful in the long-term, organizations need to focus on developing the talent that they have. That talent is the employees.

Firefighting and other emergency service work can be extremely dangerous. It unfortunately is not rare where we read in the newspaper or hear on a newscast of the death of a firefighter or other emergency service worker. Fire departments understand the risk that they face and plan for that possibility by having succession plans in place. Think about your company and what could happen. How well would your company be able to a handle loss of multiple key executives at one time? There are a number of lessons that we can learn from the emergency service community to help us prepare for the loss of key people (through attrition or tragedy).

Leaders and managers in emergency service organizations are faced with many of the same issues facing leaders in other organizations. The same tools available to you are available to emergency service organizations to help them prepare individuals for additional responsibilities. An advantage they have is that they are members of a learning organization. That's right, emergency service organizations are learning organizations. There is constant education going on.

New members are put through an orientation period and are required to successfully complete certain coursework (knowledge and competency) before they are allowed to participate at emergency scenes. Members regularly attend drills and other training sessions to hone skills and learn new ones. Officers and members that aspire to leadership courses can be found attending advanced classes on emergency service techniques, leadership, and group dynamics. These are the types of things you would expect to see in any learning organization. What is different is that emergency service organizations are also mentoring organizations. That is not to say that your business isn't a mentoring organization, but in emergency services it has always been the rule that more senior members/officers are mentors to less experienced members/officers.

The mentoring in most volunteer services is even more important because most departments have some sort of limit to the time you can spend as chief or other official rank. Knowing that you will only be the chief for two years makes succession planning a critical piece of your job. Preparing other officers to become the chief is the number two priority of the job (behind safety of the membership).

If you think about it, a new CEO every two years is probably not the healthiest change for an organization. It could lead to repeated changes in the direction the organization is heading as well as changes in other priorities. Although change can be healthy, repeated changes in direction can be confusing and costly. Keeping an eye on the VISION, having a long range plan that junior officers and members have played a part in developing, and doing succession planning are the keys to continued success of our organization in spite of the frequent changes in leadership. This is why the mission and vision statements must be able to carry the organization and gather all other activities with them. Succession planning is dependent upon reproducing the values that are espoused in the mission and or vision statements.

Our vision lets our members see that what they do is important and shows us where we are going. It is also our guide in the areas of long range planning and succession planning. Our fire department has nine elected firematic officers (A chief, two assistant chiefs, two captains, and four lieutenants). All of these officers are involved in the development of our long-range plans. That gives them some level of buy in at the onset, and helps smooth directional changes whenever we get a new chief. A new chief may make some subtle changes in direction, but he or she has already bought into the vision and direction, so wholesale direction changes are rare.

When they are necessary, all the officers have input into the changes to be made.

How do we go about succession planning? It was stated earlier that we are a mentoring organization. Our succession planning is based on mentoring from superiors, formal coursework, and experience. There are minimum requirements for formal coursework and experience for each office. The individual abilities of each person and his or her commitment to moving up both play important parts in the mentoring process. Members who aspire to be officers or who show the potential to do so are given assignments that will help prepare them to be a lieutenant (conducting a training session, running a crew at an emergency scene, etc.). A captain and an assistant chief are assigned to each lieutenant to help them with the transition from member to officer (employee to supervisor). Responsibilities assigned are used to help each officer develop the skills and experience needed to move up the line. In each position they spend part of their time learning how to perform the tasks of the next position and part of their time teaching others how to do "their current job."

On-the-job experience is a great teacher. In emergency service organizations, our emergency scenes are a mentoring event. The first arriving officer will take command of the scene (Incident Commander-IC) until relieved of the responsibility by a more senior officer. Because we are a mentoring organization, a more senior officer may allow the junior officer to continue command of the incident. They will then critique the handling of the incident as a learning experience. The decision to allow a junior officer to continue command or not is based on the complexity of the incident and how well it is being handled. (Remember, priority one is safety of the membership). This type of mentoring continues down the line at an incident. At any given time there may be an officer or experienced firefighter overseeing a less experienced firefighter or crew leader. There may be an experienced driver/pump operator overseeing the work of a new or less experienced driver.

Teaching takes place at almost all emergencies. Whenever we critique a fire or EMS call we are teaching and helping insure a successful succession plan. Whenever we mentor a junior officer, critiquing their actions at a call, we are helping assure the quality of our successor.

Job responsibilities are not confined to emergency incidents and training. Emergency service organizations are businesses. Officers must answer to a board of commissioners and/or government officials. They must deal with budgetary, personnel, and risk management issues. They often deal with the media, and must be adept at dealing with customers and suppliers for their business to be successful. These are key areas where mentoring and training are used to build a succession plan.

The training and mentoring do not stop once you become the chief. During your tenure as an officer you develop relationships with many people inside and outside the organization. Many have preceded you up the line, and are in the best position to mentor you through the issues you are now facing on a day-to-day basis. The only thing that is needed is the skill to listen. Listening is not a simple skill. It is so complex that some organizations include entire courses on it in their training program. In order to listen, many things have to happen. First, you have to put your ego on the back burner. You must put aside the idea that you are always right and just listen. Next, you must not translate what was said into your own internal language, but should listen to what is being said as it is being said. You must be willing to say, "Maybe I could learn something from that!" In succession planning for the future, we must coach the skill of listening. If someone is not a good listener something amazing will happen: the people around them will stop telling the truth. Once that happens the general will no longer get any news from the front.

As chief, you are now in a position where you spend a great deal of your time mentoring others, making sure they will be able to do your job when you are not available or serve in your place when your term is done. It is the second most important job you have, and the reason it is so important is that the most important job you have (keeping the members safe) depends on how well you do the second most important one.

How good a job is your company doing of succession planning? The following questions can help you determine where your plans need to be improved.

- Identify the key players in your organization.
- Identify the people currently in your organization that could step in and fill the role of each key player.
- Develop an action plan to improve succession planning.
- Now, have your key players repeat these exercises looking at their key players.

Firestarter 8:

Always Think Quality

It is circumstance and proper measure that give an action its character, and make it either good or bad.

—*Plutarch*

On a scale of one to ten, rate yourself on these:

- Our company knows what it wants to measure.
- Our company knows what skills it wants to shine.
- Our company is aware of areas needing improvement and has developed a plan for improvement.

The Ford Motor Company uses the phrase "Quality is Job One." We like that idea, and have just taken it to its logical conclusion: quality is job one, two, and three. As a matter of fact, you could say that quality is our only job. But how can we achieve it and maintain it?

We measure what we value. If we don't value something then we don't have to measure it. If you look at what is important to people, you will be able to tell the value they place on an item by where in their priorities this item is placed. This is why it is of vital importance to understand what business you are in and how that business can be improved by benchmarking. When you benchmark something, you give it value, have the ability to measure it and thus can improve the process. It also allows us to see when we are slipping as benchmarks degrade.

If we are in the manufacturing business, we must benchmark to make sure that our quality is consistent. If we are in the sales business, we must make sure that we are meeting our quotas. If we are in the marketing business, we will want to know if our ads are effective. There are many more things in management that are not benchmarked and thus may not be taken as seriously. If you read the mission statement of any company you must ask yourself "is everything mentioned in this mission statement benchmarked so that we can see if we are meeting our goals?" If our mission statement talks about something and we have no way to measure it we are not telling ourselves the truth about what we consider important.

There is a story that circulated on the Internet for a long time. It tells of a philosophy professor who puts a jar on his desk and fills it with large rocks. He asks the class if the jar is full. They reply that it is. He then takes out a box of smaller rocks and is able to add them to the jar around the big rocks. He asks now is it full? The class is not sure but they again say they think so. He then takes out sand and is able to do the same thing. Now the class is not sure at all what "full" means. Finally he takes out a cup of water and is able to add that too. What is the moral of the story? Put your most important tasks first and everything else will fit into your life. What are the big stones in your life?

We can also apply this lesson to business. What are the tasks that are crucial to the business? Where are the points where the customers meet the business? Where is the point at which sales or the supply chain is failing? Are we measuring this?

At the firehouse, we know that there is little distinction between large tasks and small tasks. We have learned from experience that large tasks are only successful if all of the small steps that lead up to them are done correctly. Thus benchmarking is an element that must be done every day to almost every task. Let's take a simple example.

Every truck in the firehouse is checked once a week. This check is documented on a sheet called a weekly truck check. There are benchmarks for everything on the checklist. There is a minimum amount of air in an air bottle before it is replaced. If a vehicle is below three-quarters of its capacity of the fuel tank, that tank is filled. If tires are under a certain minimum pressure, air is added. There should be a benchmark for every task at every occasion.

In addition, there are skills we benchmark. Once a year all interior firefighters must go through an obstacle course blindfolded. They are hooked up to an air bottle with a limited supply of air and must navigate this obstacle course before they run out of air. Along the way, they must climb through windows, crawl under beams, and remove and replace their air packs in order to be successful. This is done so we know that the firefighters can help in their own rescue should they become trapped within a burning building. This benchmarks a minimum competency.

Once a year we all re-certify in Cardio Pulmonary Resuscitation (CPR). We go through the steps painstakingly and exactly. We act out what we should be doing if someone needs CPR. We practice how to inflate the lungs, how to do chest compressions, what is done in what order, how many compressions must be done for each breath, and so on. We constantly practice and rehearse these skills for that time when they are needed.

When firefighters arrive at a fire we have a strict set of rules about what is done and in what order—who carries what tools, who gets the hose, who does what task. At drills these are timed and practiced over and over again until the sequence is flawless. Anything less is not acceptable.

Every task that is crucial to the success of your business should be benchmarked. If a task is not benchmarked, you don't consider it crucial, and you should evaluate why the task is performed in the first place.

It can also be beneficial for a business to benchmark your performance against others in your industry. You must measure yourself against the "best practices" in your industry. To do this you need to identify what the best practices are. In the volunteer fire service we often use paid departments as a benchmark. We are trying to accomplish the same end results, and their practices are a logical starting point for others in the industry. For example, the obstacle course that our firefighters must complete annually is adapted from the mask confidence courses used by professional firefighters. Each obstacle to be overcome was put into the course because it was responsible for the death of a firefighter in the line of duty. Being able to measure someone's ability to successfully perform a self-rescue becomes an important measure when a fire chief may need to commit those people during an emergency situation.

We can also stay at the leading edge of our industry by spending time reading industry magazines, visiting other fire departments (paid and volunteer), meeting with suppliers and manufacturers, and attending training sessions. Trade shows are a good place to stay up to date and gather new ideas. The networking that can be accomplished can give you a place to start when setting new benchmarks for your organization.

As individuals we should also be setting our own benchmarks. We strongly believe that everyone should be setting goals for themselves in order to achieve their own vision. Setting up benchmarks can help you measure how you are progressing down the road of life. Having measures will allow you to make mid-course corrections to get you where you want to go.

- What tasks are critical to the success of your business?
- What tasks are you doing that you are not benchmarking? Why?
- Are there best practices in your industry that you are not meeting? What changes can you implement to improve?
- Identify five sources of information available to you to stay abreast of the "cutting edge" in your industry.
- Identify five people that you can contact to obtain information on best practices in your industry?

Firestarter 9:

Don't Sell Yourself Short

"Destiny is not a matter of chance; it is a matter of choice. It is not something to be waited for; but rather something to be achieved."
—*William Jennings Bryan*

On a scale of one to ten, rate the accuracy of the following statements, as they pertain to your organization.

- Our company has a positive image in the minds of our customers.
- The community views us in a positive light and considers us a good neighbor.
- Our employees are proud to work for us, and are proud to show it off to the people they come in contact with.

How you manage your image can have a major impact on the success of your business. How do you overcome the perception that you are just another Enron out to make money for the chosen few at the top at the expense of the shareholders, customers, and employees? Let's take a closer look at public relations as a tool to business success by looking at the volunteer fire service.

The volunteer fire service has always been considered a group of people that were there to help you, but its members have not always enjoyed the perception of being unpaid professionals. In many communities, parts of the public sector viewed these volunteers as a group of "good old boys" that sat around drinking beer all day waiting for the whistle to blow. Once it did blow, the "boys" would be found flying down Main Street in their fire trucks, forcing the unsuspecting off the road on their way to small fires. That was an image that was formed in the minds of a lot of people based on the contact they had with their local fire department or in discussing it with others. It was an image that was not entirely false, and a difficult perception to overcome. The people that held that perception were (and still are) our customers. The vast majority will never make use of our services, but they are the people that we need to convince (read: sell) whenever we need a new fire station, fire engine, or ambulance. We need to convince these people that we are worthy of their charitable contributions, at a time when thousands of non-profits and charitable organizations are competing for shrinking dollars.

How do you overcome that type of perception? Earlier in this book we discussed our vision: "To be recognized as a leading provider of integrated emergency services." Not one mention of beer drinking good old boys in the entire statement. There are a number of good reasons why that isn't mentioned, but the number one reason is because we are not beer drinking, good old boys. We are a group of people (men and women) dedicated to serving our community, helping people in need to deal with the tragedies in their lives. There are lessons to be learned in what we did to change how we are viewed. These are actions that can be used by any organization to manage their image. All these actions can be lumped under the heading of Public Relations.

To live up to the vision statement in our fire department it became obvious that we had to have rules of conduct for our officers (leaders) and members (employees). These rules and/or guidelines on how to deal with the general public, the media, and other agencies help us stay focused. Our department has outlined these rules in our Policies and Procedures Manual and in our Standard Operating Procedures (SOPs). These are covered with all new members as part of their orientation. It covers the foreseeable circumstances we face. It gives our members guidance on how to handle certain situations before they are faced with them. Let me give you a couple of examples: We have a policy on use of intoxicants. It states that you will not respond to emergencies if you have been using any substance (alcohol, prescription or non-prescription drug, etc.) that could impair your judgment. It is simple. If you want to have drink, go ahead, but don't even think about going on an emergency call once you have. We have an SOP that covers members being approached by members of the community and/or press at an emergency scene. This is something that happens regularly, and our members know that all questions are to be referred to the Incident Commander. We feel that is important, because our ability to manage the media and flow of information is a key element in our public relations. It is one of the ways we manage our image.

We have talked a lot in this book about the things that help make our members proud, and that they are often wearing jackets, hats, and t-shirts, and driving cars that make them easily identifiable as members. The way that they conduct themselves becomes a direct reflection on us. It is no different in your organization. You should have a code of conduct or other guidelines for your employees to follow in dealing with the general public, customers, suppliers, etc.

Volunteer fire departments do a lot of good things for their community. More often than not they will go unnoticed, unless there is a system in place to capitalize on the good. For example, when we have a catastrophic fire the press is always present and we are normally shown in a positive light, but the press is seldom around when we are giving tours of the fire station to nursery school classes, or providing fire prevention education to the elementary school children, or providing potentially life saving instruction in Cardio-Pulmonary Resuscitation (CPR) to community groups. What plan does your organization have in place to work with the media to get your "story" out? Newspapers, television, and radio stations are always looking for a good story. Why not use that to help you get your story out? We have plans in place to help us work with the media in a win/win relationship. At major incidents, our Incident Command System provides for an Information Officer. This person will give regular updates to the media present to make their job easier, and hopefully keep what is reported as accurate as possible. Most everyone has seen this in action. Look at the press conferences held by Mayor Giuliani and President Bush in the aftermath of the terrorist attacks at the World Trade Center and Pentagon.

We also work with our friends in the press to get our other stories out. One story was about an elementary school student who, after attending our fire prevention education, saved her family's home (at least the kitchen) by smothering a grease fire started by her mother. Another story concerned a baby that was delivered by a National Guardsman who received his Emergency Medical Technician training as a member of our fire department. There was a story about an individual member who was recognized for saving a life. It is in an organization's interest to tell stories that present them in a good light. How are you doing that?

You can make use of the media to help you get your story out. Local newspapers and radio stations are always looking for nice stories of interest to their listeners. Having plans in place on how you will respond in the event of a disaster or catastrophic event can also help in maintaining your image as a good corporate neighbor.

There are other ways to get your story out to your customers, potential customers, and others of importance. You can and should look at developing newsletters, flyers, and promotional materials to help tell your story, but the single most important public relations tool you have is your people. Are your people actively involved in public relations for your company? What image are they presenting to your customers, shareholders, and the general public? What are you doing to make it easy for your employees to show your company in a good light? Are you a good corporate citizen? Are you providing your customers with a quality product at reasonable price? Your future success can be directly tied to the things you do in the present.

Are you and your employees meeting with community groups, serving your communities as volunteers, and being good citizens? This can play a part in how your company is viewed. Your story should include the stories of your people doing their part. There are countless opportunities for people and companies to be good citizens. Organizing a community clean up, coaching a youth sports team, volunteering at a local hospital, visiting and/or reading for the elderly at a local nursing home, or volunteering to help out your local fire department or ambulance corps are just a few ways you and your people can make a difference. It can be very rewarding and is a great way to meet new people, share your story, and hear theirs. Fire departments help build their stories by the activities of their members in much the same way your employees can. Your employees are your greatest public relations assets. Make sure they are able to tell your story, the way you want it to be told.

- Identify five people in the local media that you could approach with your story.
- Do you have contingency plans in place to help you manage a catastrophic event? When were they last updated? Have they ever been tested?
- Identify four ways that you could inform employees, customers, and the general public of positive stories about your company.

Firestarter 10:

Don't Get In Your Own Way

"It is an agile man who can stay out of his own way."

—Bob Schachat

On a scale of one to ten, rate the accuracy of the following statements, as they pertain to your organization.

- Our business rarely shoots itself in the foot.
- We have never made a mistake that cost us a customer.
- We never fail to provide the service that a customer wanted.

If you can say, "never happened" to all of the above, congratulations! But if you have, on occasion, gotten in your own way then you must think about how to stop doing that. There has been talk about "self-organizing systems" over the last decade. At key times, organizations can accomplish amazing tasks in very short periods of time. Emergencies bring this out. When there is a war, nations can suddenly manufacture thousands of tanks and planes in a short space of time, because it must be done. At times like these, nations and organizations stop getting in their own way and get the job done quickly and efficiently. Isn't this the way you would like to operate every day?

To see how this can happen in times of crisis look at what happened at the New York City Mayor's Office of Emergency Management following the terrorist attacks on the World Trade Center on September 11, 2001. The Office of Emergency Management was in the World Trade Center complex and was destroyed in the terrorist attack. On that day, there was no office, no phones and no computer network to organize the rescue efforts. That same day a site was selected along the Hudson River near 54th Street in Manhattan where the cruise ships normally dock. Several huge buildings were taken over and rapidly converted into the new Office of Emergency Management. This was done with the help of the Federal Emergency Management Agency (or FEMA). Within 36 hours of the site being set up there were 500 phone lines, hundreds of computers, thousands of cell phones, and room for over one hundred agencies to work together to solve problems. In that large cruise ship terminal there were areas for transportation agencies, law enforcement, military, public health, emergency services, a map division, a weather division, information sections, logistics sections, Red Cross and Salvation Army representatives, and sections for city, state, and federal government. Two new area codes were created and a whole computer network was set up and running. This was where Mayor Giuliani gave his nightly press conferences.

Chief McCluskey had the opportunity to work at the Logistics Station at the Office of Emergency Management during the recovery effort and had a chance to see how agencies that ordinarily did not cooperate suddenly worked well together. There was fuel when it was needed; New Jersey law enforcement and New York law enforcement worked seamlessly. The federal, city, and state emergency services interfaced without a hitch or questions about who was in charge. Suddenly, in a city where it normally takes three permits and the okay from six unions to get things done, massive jobs were getting done as soon as they were assigned. Goods and supplies poured into our warehouses from all over the world. Teams of experts flew in and assisted us. Whatever special equipment was needed, it was gotten within hours.

In those days, we did not get in our own way. There was little squabbling about territory or job title. People were doing what they needed to do to get the job done. A "company" that did not exist a week before was up and running 24 hours a day, seven days a week and doing the impossible.

This is an example of a self-organizing system. When it is time for things to get done, sometimes all we need to do is not to get in our own way.

Your business should be doing this everyday. Your business should be solving problems like this every minute it is open. But somehow you have policies or practices that actually make you the problem. Think about the story of the Mayor's Office of Emergency Management. What lessons can we learn to help us run our businesses more efficiently and effectively?

- Are their policies you could change to improve efficiency?
- How does your business get in its own way?
- How could you streamline key tasks so your employees don't stop them from being completed?

Final Thought:
Would You Be Willing To Bet Your
Life That You Are Doing A Good Job?
Firefighters Do

"The wise man is one who knows when he does not know."
—*Socrates*

On a scale of one to ten, rate the accuracy of the following statements, as they pertain to your organization.

- Is there a passion in your organization to get the job done?
- Is your team totally committed to the vision of the company?
- Is there a sense of urgency in your workplace?

Organizations have gone out of business because they lacked focus. There was little direction or energy. Often in business there was no sense of urgency. You would like your employees to be pursuing excellence at such breakneck speed that you might be tempted to ask them the question that the traffic cop asks the motorist he pulled over: "Where's the fire?" That question is actually very revealing. The cop has not pulled over a fire truck on its way to a job. He would not do that because he knows that they are on a mission. Your job is to realize that urgency in your organization every minute of every day.

If you have people who are spending the day chatting on the phone or cruising the Internet, you have not effectively communicated a sense of urgency. If you have employees that drop the ball if you don't keep after them, you have not effectively communicated that urgency. If your business is not moving at record pace, you have not effectively communicated that urgency.

There is a simple question Chief McCluskey asks at our firehouse whenever there is a dispute about some number or adherence to some Federal law. He asks the person he is talking with "Would you bet your life on that answer?" This puts things in a different light. Suddenly, the person that was arguing with such conviction and energy stops and asks him or herself if they really know what they are talking about. That question does a wonderful job of putting issues in perspective.

The reason why that question is appropriate to the fire service is that we do bet our lives on the jobs we do. If we are wrong we could be seriously injured, or worse. In the United States in 2001, there were 1.7 firefighter deaths and hundreds of injuries for every 10,000 fires. Every day, firefighters are seriously injured and every week there are reported fatalities. Just as the old adage goes, "there is nothing like an imminent hanging to focus one's mind." We could change that and say, "There is nothing like the possibility of a fire to do every job with care and attention."

The question "Would you bet your life on that answer?" asks the other person to really focus on what he or she is saying. Does he or she understand what is at stake? Does he or she understand the concepts behind the idea? Is he or she sure they can carry out what they promised? Does he or she know the difference between their own theories and the facts? Does he or she know what unintended consequences might flow from a simple decision?

While all of this may seem complicated, it comes down to two simple concepts. These are at work in great fire departments and great businesses.

They are the ideas of passion and commitment. What makes fire departments great can be summed up in those two words. Passion without commitment will never deliver on its promises. Commitment without passion will never get to the fire before the building burns down.

To really enjoy a job, you have to do something you love. If you love your work you will want to do it the best you can. To do this you need to have passion and commitment and keep those two flames alive. Volunteer fire departments have kept up high levels of passion and commitment for hundreds of years. They don't get bored, they don't slack off, and they don't go through business cycles.

In our volunteer fire department, we had a fire that went on all night in sub-zero temperatures. After being on our feet all night and being totally exhausted we went home to bed to get a little sleep before we went to our paying jobs. Shortly after we got home, the pagers sounded and the whistle started blowing for another house on fire. Even though we were exhausted and our equipment had taken a beating, that second fire was attacked with energy and focus and quickly extinguished. Is your team ready to go that extra mile?

- How could you go about lighting a fire under your team or organization so it operated with passion and commitment?
- What could you tell your employees to get them see the urgency of doing the job well?
- How could you make your workplace a place where the team wanted to achieve something great?

WOULD YOU LIKE YOUR ORGANIZATION OR TEAM
TO BURN FOR SUCCESS?

WOULD YOU LIKE TO LIGHT A FIRE
UNDER YOUR COMPANY?

If so, contact us for speaking engagements, workshops, or consultations at:

burningforsuccess.com
Or call toll free 866-347 3362

About the Authors

Chief Scott Harkins is President of Harkins Consulting Group, Inc., a firm specializing in helping clients increase profitability by finding solutions to issues that senior managers face on a daily basis. With over 20 years of progressively responsible experience in the insurance industry, Scott has successfully helped small businesses, midsize companies, and Fortune 500 corporations achieve better bottom-line results. Scott is a 29-year veteran of the fire service. Sixteen of those years were spent as a firematic officer, including seven as a Chief Officer. In addition to fighting fires, Scott proudly serves the fire department as a NY State Certified Emergency Medical Technician (EMT-D), and NAUI Certified Rapid Deployment Underwater Rescue Diver. A degreed fire protection and safety engineer, Scott is a Certified Safety Professional (CSP) and a Chartered Property and Casualty Underwriter (CPCU). He is a past member of the Board of Directors and Chairman of the Electrical Committee of the New York Board of Fire Underwriters, is a member of the American Society of Safety Engineers, the Society of CPCU, the National Fire Protection Association, and the Board of Certified Safety Professionals.

Scott can be reached at:
chiefscott@burningforsuccess.com

"Dr. Frank" McCluskey is an Ex-Chief in the Mahopac Falls Volunteer Fire Department and Dean of Online Learning at Mercy College in New York. He has taught courses in both leadership and philosophy. Dr. Frank has a Ph.D. from the New School for Social Research and was a Post Doctoral Fellow at Yale University. He has published dozens of articles. He has led workshops in leadership, motivation, management, and applied philosophy with hundreds of employees from Fortune 500 companies. He has spoken all over the world and appeared on CBS, Broadcast New York, the Canadian Broadcasting Network, and on the PBS Business Channel with Joan Lunden and Dave Barry. His double life as a firefighter and philosopher was the subject of a story in the New York Times. He is also the author of *Thoughts on Fire: Life Lessons of a Volunteer Firefighter.*

Frank can be reached at:
drfrank@burningforsuccess.com

Index of Terms

We acknowledge that some of the terms and acronyms used in this book are unique to emergency services. We have tried to explain these terms and acronyms as they were used and now provide this index as a guide for the reader.

CBT: Competency Based Training
CPR: Cardio-Pulmonary Resuscitation
EMS: Emergency Medical Services
EMT: Emergency Medical Technician
ESO: Emergency Service Organization
FAST: Firefighter Assistance and Survival Team
IC: Incident Commander
ICS: Incident Command System
PFD: Personal Flotation Device
SCBA: Self-Contained Breathing Apparatus
SCUBA: Self-Contained Underwater Breathing Apparatus

0-595-24012-7